Fire Hazards in Industry

Fire Hazards in Industry

Norman Thomson

BUTTERWORTH
HEINEMANN

OXFORD AUCKLAND BOSTON JOHANNESBURG MELBOURNE NEW DELHI

Butterworth-Heinemann
Linacre House, Jordan Hill, Oxford OX2 8DP
225 Wildwood Avenue, Woburn, MA 01801-2041
A division of Reed Educational and Professional Publishing Ltd

A member of the Reed Elsevier plc group

First published 2002

British Library Cataloguing in Publication Data

Thomson, Norman
Fire hazards in industry
1. Fire prevention – Law and legislation 2. Fire risk assessment 3. Fire prevention – Management
I. Title
658.4'77

Library of Congress Cataloguing in Publication Data
A catalogue record for this book is available from the Library of Congress

ISBN 0 7506 5321 3

Typeset by Laser Words Pvt. Ltd., Chennai, India
Printed in Great Britain by Biddles Ltd, www.Biddles.co.uk

Contents

Introduction

'Fire spreads through factory premises'
'Two employees are killed when gas cylinder explodes'
'126 jobs lost as fire-hit company closes'
'Chemical company fails to recover after major fire'

Unfortunately, these headlines are all too common. In the United Kingdom, fire brigades attend more than 40 000 fires in workplaces every year. These fires kill more than 30 people per year and injure almost 3000 people. In addition, insurance claims for fire damage in workplaces amount to an average of £10 million per week.

Fire kills, injures and causes damage to property, and it can also have significant effects upon the future of companies. It is estimated that 60% of businesses that have suffered serious fire damage fail to recover and cease trading within five years of the initial incident.

The Association of British Insurers released a press statement in the late 1990s, which stated that:

> Britain's businesses and their insurers are being hit by a major rise in the cost of fire damage, with losses costing over £1.6 million every day. In 1998, fire damage claims cost insurers £600 million, up 22% on the previous year to their highest level since 1992. The combined cost of fire damage and business interruption losses, at £808 million, rose by 13% on the previous year.

Although this was a stark message, an even more disturbing message was perhaps contained within a further statement:

> The current level of commercial fire losses is very disturbing and needs to be tackled. These losses are unsustainable and can only put pressure on premium rates unless there is a significant improvement during the year. Businesses must recognize the urgent need to reduce the risk from fire, and make this a priority for this year.

Although accidental fires are on the increase, one of the most concerning categories of fires are those related to arson. It is estimated that fires that are started deliberately account for 40% of workplace fires. In some sectors, this figure is significantly higher. For example, in the category of recreational- and cultural-type premises, 66% of fires are started deliberately.

The risk from fire is, therefore, an important area for employers to control and manage properly. Generally, there are three main reasons for managing fire risks.

Firstly, there is a moral duty on employers to provide a safe workplace that is free from risks to health. Employees perform better at work and enjoy their working environment when they can see that their safety has been considered and that steps have been implemented to ensure that risk is controlled. In this age of modern health and safety legislation, it is not considered morally acceptable to put people at risk unless under special circumstances, e.g. emergency incidents.

Secondly, from an economic point of view, fire safety management is also important. As previously stated, fires cost significant sums of money. A common belief is that most of the damage that results from fire can be recovered from insurance companies. Although not relating specifically to fire incidents, the Health and Safety Executive carried out a study in the 1990s, whereby the true cost of accidents was investigated. As an average, it was calculated that, should an incident occur, only 8% of the loss would be recovered through insurance.

Direct costs such as property damage and personal injury are definitely recovered from insurers. However, it is the hidden costs, or the indirect costs, that are important. These costs include administration costs, loss of production, loss of orders, bad publicity as a result of accidents and possible litigation. These indirect losses may have a detrimental effect on business recovery after a serious fire. In fact, more than half of the companies experiencing a serious fire will fail to recover enough to continue trading. This figure is actually higher for small to medium-sized enterprises.

Although economic costs are important, a final reason for managing fire safety is because it is a legislative requirement. There are two main items of legislation that relate directly to fire precautions. These are the Fire Precautions Act 1971 and the Fire Precautions (Workplace) Regulations 1997 (as amended). Both of these are enforced by the local fire authorities, with the exception of Crown-occupied premises where enforcement is carried out by the Fire Services Inspectorate of the Home Departments.

The Fire Precautions Act deals with general fire precautions such as:

- Means of detecting a fire and giving warning in the event of a fire.
- The provision of a suitable means of escape from premises that come under the Act.
- The provision of suitable fire-fighting equipment, including fire extinguishers, hose reels and sprinkler systems.
- Training of staff in relation to fire safety arrangements, e.g. evacuation procedures, fire alarm systems, fire-fighting and generally awareness of fire hazards.

In addition to the Act, the Fire Precautions (Workplace) Regulations require employers to carry out fire risk assessments. Assessing the risk from fire involves the following stages:

- Stage one – identification of fire hazards, including sources of ignition, sources of fuel and work activities that may present a fire hazard.
- Stage two – identification of the number, location and type of person at risk from fire, including employees, contractors, visitors and members of the public.
- Stage three – evaluation of the risk from fire hazards. This stage involves evaluating the existing arrangements for controlling fire risks, such as

control of ignition sources and sources of fuel, fire detection, means of escape, maintenance of fire controls and training of employees. Where these existing arrangements are sufficient, no further action is required. However, where the existing controls are found to be inadequate, further control measures need to be identified, implemented and maintained.
- Stage four – recording the findings of the fire risk assessment. This may be done using a paper-based system or by electronic means. Also, employees and others who may be affected by the fire risks need to be consulted, made aware of the risks and trained in the measures required to control the risks.
- Stage five – reviewing the assessment when significant changes occur to the process or after a fixed period of time.

The Fire Precautions (Workplace) Regulations place a duty on the employer to carry out fire risk assessments. This is in contrast to the Fire Precautions Act, which requires owners or occupiers of the premises to put in place suitable means of preventing and controlling fire. It is, therefore, the responsibility of the employer to ensure that suitable and sufficient assessments are carried out and that resources are allocated to ensure that control measures are implemented.

The role of the local fire authority will be to periodically request that assessments are available for their inspection and may also require access to the premises for a physical inspection of the workplace. The fire authority will request fire risk assessments routinely, depending on the type of premises occupied, the number of persons at work and whether there is a history of fire incidents in the premises or within the area.

Another important item of legislation that applies to fire safety is the Health and Safety at Work Act 1974 and the associated regulations made under the Act. Matters falling within the scope of the Act and its associated regulations include:

- The storage of flammable materials.
- The control of flammable vapours.
- The control and storage of highly flammable liquids.
- The maintenance of equipment, plant and machinery.
- General housekeeping arrangements.
- Safe systems of work.
- Training of employees and others.

The Management of Health and Safety at Work Regulations, made under the 1974 Health and Safety at Work Act, which were introduced in 1992 and later amended in 1999, also requires employers to carry out fire risk assessments of their workplaces. Between them, the Fire Precautions (Workplace) Regulations and the Management of Health and Safety at Work Regulations require employers to carry out fire risk assessment, control the risks by implementing and maintaining suitable controls and to inform employees of the findings of the assessment. In addition, employers must make the following provisions:

- Nominate specific people to undertake any special roles in emergency situations, e.g. fire wardens, fire evacuation assistants. An emergency plan is required for evacuation purposes, which should include details of specific roles.
- Consultation with employees or their representatives regarding specific risks in relation to fire safety.
- Information to other parties who may be affected by fire risks within the employer's premises, e.g. employers of people in adjoining premises, contractors and members of the public or visitors.
- Establish a suitable means of contacting the emergency services.
- Liaise with the local fire authority should structural changes occur to the workplace.

Although fire safety legislation may at first seem complicated, it is really quite straightforward. Once the concept of fire risk assessment is understood, it is not particularly difficult to carry out assessments and to implement and maintain control measures. Self-regulation, which is a concept of modern health and safety legislation, offers significant benefits to employers and employees. Self-regulation offers a great opportunity to involve employees at all levels in the risk assessment process, which can often result in a more broad-minded approach to solving problems of fire risk and will also help to ensure that all parties take ownership of health and safety issues.

Furthermore, one of the major benefits that complying with such legislation provides is that the business can identify ways to protect its assets – including people, stock and processes – which in turn will ensure that the business continues in a manner whereby risk is controlled. Carrying out fire risk assessments not only ensures compliance with the law but also ensures that a company does not fall within the staggering statistics quoted at the beginning of this chapter, especially those concerning the high level of businesses that fail to recover when a fire does occur.

The fact that fire safety management is important was summed up by the Director General of the Association of British Insurers, Mark Boleat:

> Loss prevention remains the key to reducing commercial losses. More than ever it is essential for businesses to have in place comprehensive loss prevention and business recovery programmes. Insurers will increasingly be looking for greater emphasis on risk management to bring about a fall in the current high level of theft, fire and weather-related commercial property claims.

If more than half of the businesses that suffer fire collapse economically within five years of the incident, fire safety management is surely a worthwhile part of the overall approach to business management.

1 Fire Safety Legislation

ROSE AND CROWN HOTEL FIRE

On 26 December, 1969, 11 people were killed in a tragic fire at the Rose and Crown Hotel, Saffron Walden, Essex. It is thought that the fire started in a ground floor television lounge, probably from a faulty television set.

The fire was first detected at approximately 1.30 a.m., when two guests from the first floor noticed smoke and went downstairs to investigate. As they left the building, the couple notified the fire brigade about the fire. By the time the fire brigade arrived on the scene, the fire had developed rapidly throughout the building.

Of the 33 people staying in the hotel that night, 12 were rescued immediately by ladders, three jumped to safety, two walked out of the building and five others were assisted from the building by local residents using ladders. Unfortunately, 11 people did not manage to escape from the intense heat and smoke and their bodies were later recovered by the fire brigade.

All fatalities were found on the second and third floors of the hotel. After the fire, evidence revealed that those who died did so from the early surge of heat and gases during the rapid build-up of fire. This build-up of heat, smoke and fumes was probably enhanced by the opening of doors and windows by other residents to make their own way to safety.

The only mechanism that effectively stops heat and smoke travelling through premises such as the Rose and Crown are fire doors, which are made from fire-resisting material and fitted with self-closing devices. Although there were some fire doors enclosing the internal staircase of the hotel, the mechanisms of many doors were reported to have failed, allowing them to remain open and allowing deadly gas and smoke to engulf the staircase and pass to all other parts of the premises.

Another problem that contributed to the high death toll was the fact that the fire alarm failed to operate properly and, therefore, failed to warn other residents of the initial outbreak. It was later revealed that the alarm circuit wiring had been destroyed by the fire, resulting in the alarm, which was initially activated by one of the residents, failing to continue sounding throughout the fire.

The fire at the Rose and Crown was not the only serious hotel fire during the late 1960s. Several other fires similar in nature to the fire at Saffron Walden occurred during that time and many lives were lost as a consequence. However, it was the Rose and Crown fire that finally prompted the Government of the day to take serious action as far as fire safety legislation was concerned.

THE FIRE PRECAUTIONS ACT 1971

The Government decided that future fire safety legislation should be flexible and adaptable. Previous fire safety legislation was very prescriptive in nature, which quickly became out of date and therefore ineffective. A more flexible approach was needed, as was the need for adaptability. It was thought necessary to have an approach that allowed premises to be 'designated' if it were considered necessary to improve their standards.

The Fire Precautions Act 1971 was brought into UK legislation and was designated an 'enabling' act. This meant that the Secretary of State could designate, by means of statutory instrument, premises to which the new act could be applied. Early use was made of this new adaptable legislation when, in 1972, the Fire Precautions (Hotels and Boarding Houses) Order was added to statutory law. This meant that the Fire Precautions Act could be applied to certain hotels and boarding houses.

A further statutory order was made several years later. The Fire Precautions (Factories, Offices, Shops and Railway Premises) Order 1989 brought fire safety provisions, previously contained within the Factories Act and the Offices, Shops and Railway Premises Act, within the scope of the 1971 Act.

The main requirement of the Fire Precautions Act 1971 is that occupiers of designated premises are required to apply to the local fire authority for a fire certificate. In the case of premises where there are multiple occupiers, it is the responsibility of the owner of the premises to apply for a fire certificate.

The following criteria must be fulfilled before a certificate is required.

The Fire Precautions (Hotels and Boarding Houses) Order 1972

A fire certificate is required where there is:

- sleeping accommodation for more than six people, whether they be staff or guests, or;
- sleeping accommodation above the first floor or below the ground floor level of the building, whether for staff or for guests.

The Fire Precautions (Factories, Offices, Shops and Railway Premises) Order 1989

A fire certificate is required where:

- more than 20 persons are at work at any one time, or;
- more than 10 persons are at work elsewhere other than on the ground floor, or;
- highly flammable substances or explosives are stored or used in, or under, the premises.

When counting employees, it must be remembered that all employees must be counted, regardless of whether they work for a single company or not. For example, suppose that three companies occupy a single-storey building, each company employing 10 employees. In terms of the qualifying requirements of the legislation, each company employs less than 20 persons; therefore, a certificate is not required. However, the total number of people employed within the premises exceeds 20, which means that the premises require a fire certificate. In this case, it would be the responsibility of the owner of the premises to apply for a fire certificate from the local fire authority.

FIRE CERTIFICATES

A fire certificate is a document that requires the owner or occupier of the premises to ensure that effective measures are taken to comply with the content of the certificate. A fire certificate will specify the following information:

- Use of the premises.
- The means of escape provided for use in the event of fire, e.g. escape routes, fire exits and fire doors.
- The means of ensuring that the means of escape can be safely used at all times, e.g. fire exit signs, self-closing devices on fire doors and emergency lighting.
- The means for fighting a fire, for use by persons occupying the premises, e.g. fire extinguishers, fire blankets and hose reels.
- The means of raising the alarm in the event of fire, e.g. fire alarms.

The fire certificate may also specify arrangements to be made for the storage and use of explosives and/or highly flammable substances that are stored or used within the premises.

In addition, a fire certificate may specify the following:

- Maintenance of the means of escape and the requirement for keeping it clear of obstruction.
- Maintenance of fire-fighting equipment.
- Training of employees in the action to take in the event of fire, which may include fire-fighting, evacuation, raising the alarm and dealing with non-mobile persons.
- Limitation of the numbers allowed into the premise at any one time.
- Any other relevant fire precautions.

APPLICATION FOR A FIRE CERTIFICATE

The first stage of a certificate application is for the owner or occupier of the premises to complete a form (FP1 Rev), which is available from the local fire authority. Following receipt of the application, the authority may ask the owner or occupier to provide floor plans of the premises.

While the application is pending, it is the owner's or occupier's responsibility to ensure that they comply with their 'interim duties'. These include the following:

- Ensuring that any existing means of escape in case of fire can be safely and effectively used whenever people are on the premises.
- Any existing fire-fighting equipment with which the premises are provided is maintained in efficient working order.
- All employees on the premises receive instructions or training in what to do in the event of fire.

The second stage of the application is for the fire authority to decide whether an exemption is to be made. In order to do this, the authority must first inspect the premises. The inspector will take into consideration the means of escape from the premises, any fire-fighting equipment and any fire warning system that may be installed. In addition, the inspector will assess the type of work that is carried out and the number of people who may be present within the premises. A further consideration will be given to the type of construction, e.g. modern construction or traditional, concrete or mainly timber.

The next stage of the process is for the fire authority to:

- issue an exemption from the requirement to have a fire certificate, or;
- issue a fire certificate if the authority believe that all necessary precautions have been taken, e.g. adequate means of escape, fire-fighting equipment and adequate fire warning, or;
- issue a notice that will indicate the steps that are necessary in order to bring the premises up to an appropriate standard.

If the fire authority decide that further steps are necessary, they will give the occupier or owner a suitable time period, after which time a reinspection will take place. If, after the reinspection, the fire inspector is satisfied that fire precautions are adequate and meet the standard, he or she will then issue a fire certificate.

An owner or occupier of premises must ensure that the local fire authority is informed immediately if any structural alterations or changes that may affect the means of escape are planned. In addition, if a change of use occurs, this should also be communicated to the fire authority.

CHANGES TO THE FIRE PRECAUTIONS ACT 1971

To continue with an explanation of the evolution of fire safety legislation in the United Kingdom, it is appropriate to pause at this stage and reflect upon another major fire that resulted in significant changes being introduced into the structure of fire safety legislation.

On 11 May, 1985, a capacity crowd had gathered to watch a football match at Bradford City Football Stadium. The match was stopped after a relatively small fire was noticed in the wooden spectator stand. However, the small fire quickly escalated into one of the worst sports ground disasters ever seen in the United Kingdom.

Ten minutes after the initial outbreak, the scene at the football stadium, witnessed by millions of television viewers, was one of devastation. The fire, which started under the stand where paper rubbish had been allowed to accumulate over a period of many years, quickly engulfed the wooden stand. Hundreds of people fled onto the pitch, many scrambling over other people to make their escape. However, for 56 people, escape was impossible. Many bodies of those who had been killed were found piled up against locked exit barriers behind the stand.

THE FIRE SAFETY AND SAFETY OF PLACES OF SPORTS ACT 1987

Following the fatal fire, a Government inquiry was set up to examine the operation of the legislation in force at that time, which consisted mainly of the Safety of Sports Grounds Act 1975. As a result of the inquiry, a new item of legislation was introduced, The Fire Safety and Safety of Places of Sports Act 1987. The new legislation is important because it contained sections that made three significant amendments to the Fire Precautions Act 1971.

The first of these amendments was to allow fire authorities to relax the requirement for premises, previously designated under the 1971 act, to be provided with a fire certificate. The introduction of the new legislation had far-reaching implications, which gave the fire authority the power to grant exemption for a fire certificate if they were of the opinion that the premises was of low fire risk to those who occupied it.

Many factors have to be taken into consideration by the fire authority when making this decision. The decision by the fire authority has to be based on the 'risk', and not necessarily on the actual fire precautions within the premises. This resulted in a much less prescriptive form of fire precautions. The concept of judgement based upon the risk was later developed into further fire safety legislation.

The second amendment made it mandatory for owners or occupiers, of premises that were exempt from having a fire certificate, to ensure that they provide adequate means of escape and adequate means of fighting a fire.

The final amendment made to the Fire Precautions Act 1971 gave the fire authority the power to issue Improvement and Prohibition Notices.

IMPROVEMENT NOTICES

An Improvement Notice is a formal notice issued by the fire authority when it believes that an owner or occupier of premises, which are not required to have a fire certificate (designated as a Section 9A premise), has failed to provide adequate means of escape and/or adequate means of fire-fighting.

The notice issued should:

- state that the fire authority consider that the 'Act' has been contravened, and;
- specify the action required to remedy the contravention, and;
- specify a time period for remedying the contravention.

An owner or occupier who has received an improvement notice has the right to appeal within 21 days of being served that notice. Appealing has the effect of suspending the notice until after the hearing, which may then cancel the notice or reinstate it for remedial action to take place. Appeals are brought before Industrial Tribunals for final decision.

PROHIBITION NOTICES

A prohibition notice is a formal notice issued by the fire authority when it believes that a serious risk to life exists should a fire occur.

The notice issued should:

- state that the fire authority consider that the use of the premises involves or will involve a risk to persons on the premises in the event of a fire that is so serious that the use of the premises should be prohibited or restricted, and;
- specify the matters giving rise to that risk, and;
- direct that the use of the premises is prohibited or restricted.

A Prohibition Notice may also include the action necessary to remedy the problem. An owner or occupier who has received a prohibition notice has the right to appeal within 21 days. However, bringing an appeal will not automatically suspend the notice; this can only be done by the appeal court.

SPECIAL PREMISES

Although premises designated under the Fire Precautions Act 1971 may require a fire certificate, or alternatively be granted exemption by the fire authority, there is another category of premises that is worth mentioning. These are special premises, which involve the use of hazardous processes or the manufacture or storage of hazardous materials, which present a significant risk to employees and the general public. These premises do not come under the requirements of the Fire Precautions Act and are not subject to enforcement by the fire authority. Instead, these premises are required to comply with the Fire Certificate (Special Premises) Regulations 1976 and come under the control of the Health and Safety Executive (HSE).

GENERAL HEALTH AND SAFETY LEGISLATION

Similar to fire safety legislation, general health and safety legislation also underwent an evolutionary process. In 1967, the Government published a consultative document with a view to consolidating in one act the health and safety requirements of several existing acts, e.g. the Factories Act, and Offices, Shops and Railway Premises Act. However, even in the early years of this consolidation process, doubts existed as to the feasibility of such a project.

In June 1970, the Secretary of State for Employment and Productivity established a Royal Commission under the Chairmanship of Lord Robens to look at the whole issue of health and safety in the United Kingdom and to recommend appropriate changes to the system. The Robens report recommended that the 'haphazard mass of ill-assorted and intricate detail' should be replaced with a much more 'comprehensive and orderly set of provisions under a new enabling Act'. The report went on to say, 'The new Act should be supported by regulations and by non-statutory codes of practice'.

The foregoing new enabling Act was brought onto the statute in 1974, under the title of 'The Health and Safety at Work etc. Act' and provided a framework for future health and safety regulations and guidance. Although the Act did not specifically mention fire safety, future developments were to bring about major changes to fire safety legislation.

The main achievements of the Health and Safety at Work Act were as follows:

- The facility to develop and install secondary legislation as regulations. Since 1974, there have been many health and safety regulations introduced, which provide greater detail than the Act itself on subjects such as noise at work, asbestos at work, manual handling operations, control of chemicals hazardous to health, construction and fire safety.
- Bringing together various inspectorates into one organization, The HSE. The HSE is under the supervision of the Health and Safety Commission (HSC), which consists of representatives from industry, commerce and trade unions. The HSC is now a world leader in research, training and information on a wide range of health and safety issues.
- Providing inspectors with powers to issue improvement notices and prohibition notices.
- The provision for inspectors to prosecute. The maximum fine that can be imposed by the Magistrates' Courts is currently £20 000 for breaches of certain sections of the 1974 Act and £5000 for breaches of most other sections. Where cases are heard in Crown Courts, there is no limit on the fine that can be imposed. In addition to fines, courts can sentence people to imprisonment for serious offences.
- The provision for appointing safety representatives and safety committees and the rights of safety representatives to information and consultation.

GENERAL DUTIES OF EMPLOYERS

One of the most important sections of the Health and Safety at Work Act is Section 2, 'General Duties of Employers'.

Section 2(1) 'It shall be the duty of every employer to ensure, so far as is reasonably practicable, the health, safety and welfare of all his employees'.

Section 2(2) provides more detail to the above general statement and requires an employer to provide:

- the maintenance of safe plant and systems of work
- the safe use, handling, storage and transport of articles and substances
- information, instruction, training and supervision on the risks to health arising from workplace activities
- safe places of work and safe means of access and egress
- a safe working environment and adequate facilities and welfare arrangements.

Section 2(3) provides that an employer with five or more employees must prepare a written health and safety policy statement. The statement should confirm that the employer will take responsibility for the health and safety of employees and others affected by his activities. In addition, the employer must put in place an organization for health and safety and must make arrangements for identifying and controlling risks. The general policy and the arrangements for controlling risks should be brought to the attention of all employees. The employer should consult with employees and, in some cases, others who may be affected by his activities while developing policies and arrangements.

THE 1987 SINGLE EUROPEAN ACT

Interestingly, the Robens Report was published in the same year that Britain entered the common market through the 'European Communities Act 1972'. Some time later, in 1987, Britain signed the 'Single European Act', which amended the earlier 'Treaty of Rome'. The Single European Act permitted the European community to introduce minimum standards for the health and safety of workers within the community. It effectively meant that, if a vote was successful in the European Parliament, a Directive could be issued to all member states, requiring them to implement the Directive into its own health and safety legal system. This has resulted in a major revolution in health and safety legislation in this country, including some far-reaching implications for fire safety.

THE FRAMEWORK DIRECTIVE

The first and most important of the post-1987 Directives was the Framework Directive. This Directive set out the strategy for all future Directives, which generally requires employers to:

- avoid risks to health and safety
- evaluate risks that cannot be avoided

- combat risks at source
- adapt the work to the individual
- adapt to technical progress
- replace the dangerous by the non-dangerous or the less dangerous
- develop an overall protection policy
- provide information and instruction to employees.

The Framework Directive was followed by six other 'daughter' Directives, all of which were implemented by 1993. In the United Kingdom, these Directives were implemented through the provision of the Health and Safety at Work Act as the following regulations:

- Management of Health and Safety at Work Regulations 1992 (superceded in 1999)
- Workplace (Health, Safety and Welfare) Regulations 1992
- Personal Protective Equipment at Work Regulations 1992
- Manual Handling Operations Regulations 1992
- Health and Safety (Display Screen Equipment) Regulations 1992
- Provision and Use of Work Equipment Regulations 1992 (superceded in 1998).

The Management of Health and Safety at Work Regulations 1999

The main requirements of the Management of Health and Safety at Work Regulations are as follows.

Risk Assessment

All employers and self-employed people must assess the risks to workers and any others who may be affected by their work or business. All employers should carry out a systematic general examination on the effect of their undertaking, their work activities and the condition of their premises in relation to the risk to employees and others. Those who employ more than five persons should record the significant findings of the risk assessment.

In addition to general risk assessments, employers should carry out fire risk assessments. These should include the identification of fire hazards within the workplace, the number and type of persons likely to be affected by those hazards and the evaluation of risk from those hazards.

Risk assessments should be used as a continuous improvement process, which should involve action planning and resource allocation for controlling risk, both in the immediate term and the longer term.

Principles of Prevention

Employers and the self-employed need to introduce preventative and protective arrangements to control risks (and fire risks) identified by the risk assessment. As far as possible, risks should be eliminated completely. However, in many circumstances this would be impractical; therefore, some degree of cost and benefit should be applied.

Health and Safety Arrangements

Employers must have arrangements in place to ensure that there is a continuous improvement process for health and safety. Employers must ensure that their management system takes into account the following:

- Plan – a health and safety policy should identify how a company or organization will deal with health and safety issues.
- Organize – by allocating responsibilities for health and safety and for the implementation of the health and safety policy.
- Control – through the use of performance standards and the development of appropriate management systems.
- Monitor – through regular health and safety inspections and audits.
- Review – through regular health and safety meetings and safety committees.

Health and Safety Assistance

Employers must ensure that they have access to competent help in applying the provisions of health and safety legislation, including fire safety law. Assistance should be sought from competent health and safety practitioners to help in the development of policies and procedures and applying principles of prevention and protection.

Procedures for Dealing with Serious and Imminent Danger

Employers should establish procedures for any worker to follow if situations present serious or imminent danger, e.g. a fire. Procedures should identify the following:

- The nature of the possible danger and the risk that may be present.
- Additional procedures as are necessary to control the situation.
- Allocation of additional responsibilities.

Information for Employees and Others

Employers should ensure that employees are provided with comprehensive and adequate information regarding the risks that they face in relation to their health and safety. In addition, information may have to be provided for other persons who may be affected by the employer's work activities, e.g. members of the public and visitors to the workplace.

Workplace (Health, Safety and Welfare) Regulations 1992

These regulations require employers to maintain the workplace in a safe condition. The regulations refer to standards of lighting, temperature control, ventilation and for the provision of facilities such as lockers and toilets.

Another important requirement under these regulations is for standards of housekeeping to be adequate to maintain the workplace in a safe condition. Poor housekeeping is one of the most common causes of fires in the workplace.

Personal Protective Equipment at Work Regulations 1992

These regulations require employers to assess the need for personal protective equipment and to provide such equipment that helps to control risks to health. As far as fire safety is concerned, employers should consider the need for any item of personal protection that may be required should fire break out in the workplace. For example, if a procedure were to be developed whereby emergency shutdown of complex chemical plant was required in emergency situations, the employer may have to provide adequate means of personal protection for employees who are likely to be involved in such work, e.g. gloves, special suits and, in some cases, respirators and breathing apparatus.

Manual Handling Operations Regulations 1992

These regulations relate to tasks involving physical movement of loads.

These regulations do not have many implications as far as fire safety is concerned; however, they should be borne in mind when locating fire-fighting equipment. For example, if employees were expected to fight a small fire in the workplace, an assessment of the manual handling implications of any equipment that has been provided may have to be undertaken.

Health and Safety (Display Screen Equipment) Regulations 1992

These regulations require employers to carry out an analysis of computer workstations, mainly in terms of ergonomic implications for users of such equipment.

Provision and Use of Work Equipment Regulations 1992 (Superceded in 1998)

These regulations are important in terms of fire safety. Employers are required to:

- assess the suitability of items of equipment used within the workplace and those used by employees in work-related activities out of the workplace;
- ensure that equipment has been designed to appropriate standards;
- ensure that equipment is regularly tested in accordance with the manufacturer's instructions; and
- ensure that employees know how to operate the equipment safely.

A common cause of fires in the workplace is due to poorly maintained equipment and machinery. Compliance with these regulations should ensure that all items of equipment that may present a fire risk have been regularly inspected and tested.

THE FIRE PRECAUTIONS (WORKPLACE) REGULATIONS 1997, AS AMENDED 1999

As a result of another European Directive, the above regulations were introduced into health and safety legislation in the United Kingdom in 1997. The main purpose of the regulations was to introduce the concept of self-compliance for fire safety. In keeping with all other health and safety legislation, the 'Workplace' regulations require employers to carry out fire risk assessments and to control risks identified as a result of those assessments.

Before the Fire Precautions (Workplace) Regulations 1997 were introduced, owners and occupiers of premises could rely on the enforcing authority (usually the local fire authority) to inspect premises and produce a report on their findings. All the employers had to do was to comply with the fire authority's requirements and they could be sure that full legislative compliance had been met.

However, with the introduction of the above regulations, the situation is now vastly different. Employers (and not owners or occupiers) must assess the risk of fire within the premises arising from work activities. The assessment must take into consideration the existing fire precautions that exist within the premises and an evaluation made as to their effectiveness in relation to the risk presented.

In addition to the requirement to carry out a comprehensive fire risk assessment, the other main requirements of the regulations include the following.

Application of the Regulations

The regulations apply to 'workplaces', which means that any premises or part of premises (excluding domestic premises) used for an employer's undertaking. Workplaces occupied by persons who are self-employed, and who do not employ others, are also exempt from the regulations.

Information to Employees

Employers must ensure that employees are provided with adequate information on:

- the risks to their safety identified by the fire risk assessment
- the preventive and protective measures
- allocation of responsibilities for dealing with the above.

Cooperation and Coordination

If workplaces are shared by other businesses, the employer has a responsibility to make all other employers aware of any risks found that might affect employees.

Enforcement

Although the Fire Precautions (Workplace) Regulations 1997 have been made under the provisions of the Health and Safety at Work Act 1974, the authority responsible for enforcing the regulations is the local fire authority.

Offences

Offences made under the Fire Precautions (Workplace) Regulations will carry the same penalties as set out by the Health and Safety at Work Act.

Prohibition Notices

The local fire authority can serve a prohibition notice, which can prohibit or restrict the use of a workplace until the risk to employees has been reduced.

Improvement Notices

The local fire authority can serve improvement notices requiring an employer to improve the fire precautions within a workplace. The authority may issue a notice if they consider that a particular regulation has been breached. The notice must:

- specify the steps required to remedy the breach;
- require those steps to be taken within a given time; and
- provide details of the appeals procedure relating to enforcement notices.

Exemptions

The following premises are exempt from the Fire Precautions (Workplace) Regulations 1997:

- Workplaces used only by the self-employed
- Private dwellings
- Construction sites
- Ships, within the meaning of the Merchant Shipping Act 1995
- Mines, excluding surface buildings
- Offshore installations
- Agriculture or forestry land workplaces situated away from the main buildings.

THE FUTURE

In 1998, the Court of Appeal made a significant decision, which is likely to have major implications for courts awarding penalties for breaches of health and safety law. In the case of Howe and Son (Engineers) Ltd, the Court of Appeal said that the magistrates should consider carefully before hearing health and safety cases, particularly if they feel that by accepting jurisdiction they may be limiting the fine imposed. Magistrates can only fine a company a maximum of £20 000 for each of the general duties of the Health and Safety at Work Act and a maximum of £5000 for most other offences. Owing to the decision made by the Court of Appeal, it is very likely that we will see many more cases being referred to Crown Courts, where unlimited fines may be imposed.

In addition to larger fines, a move is being made towards introducing the offence of 'corporate killing'. In 1996, the Law Commission recommended changes to the law of manslaughter, whereby a corporation would be found guilty of the offence if:

- a management failure by the corporation was the cause of a person's death, and;
- that failure constituted conduct that was short of conduct to be reasonably expected of such a corporation.

In May 2000, the Government announced its intention to introduce the above offence.

This change to the way in which health and safety is enforced is likely to mean that more offences will come before courts for trial. It may also mean that the more serious offences are dealt with on a more personal level. Employers should be held accountable for their actions, or lack of action, in terms of securing the health and safety of their employees. Current legislation and future legislation may help to ensure that this accountability is fully met.

REFERENCES

Wood, P. (1970) Fire – Rose and Crown Hotel, *In Attendance Magazine, issue number unknown.*

Hendy, J. and Ford, M. (1998) *Redgraves Health and Safety*, Butterworths.

HSE Books (1999) *Fire Safety – an employers guide*, HMSO.

2 Fire Risk Assessment

HICKSON AND WELCH

In September 1992, a major explosion occurred at the premises of Hickson and Welch in West Yorkshire, killing four people and injuring many more. The Health and Safety executive brought the case to court and the company was fined £250 000, with costs of approximately £150 000. Mr. justice Holland, who presided over the case, said,

> There was no safe system of work, none such was maintained, notwithstanding the hazards that were to be perceived, the hazards that were there as a potential. This was not a casual breach of an employer's duty, but a plain gap in the employer's management which should never have occurred.

At approximately 1.20 p.m. on 21 September, 1992, a jet of flame erupted from a batch still, which then struck a four-storey building occupied by more than 60 people. Fortunately, most of them escaped from the office building, with the exception of one young girl who died because of smoke inhalation.

Hickson and Welch employed more than 900 people at the factory, which manufactured organic chemicals and specialized in the manufacture of aromatic compounds such as nitrotoluenes. The plant had a capacity to produce 60 tonnes per day.

Since its installation in 1961, the batch still had never been opened for cleaning. As a result, sludge had accumulated in the still, resulting in a degradation of the process. One operator recorded the depth of the sludge to be 34 cm.

On the morning of 21 September, it was decided that the sludge had to be removed. A skip was ordered for disposal of the sludge, some steps were removed to provide easier access and scaffold was erected.

Steam was applied to soften the sludge; however, it was estimated that a problem may occur if the temperature inside the vessel exceeded 90 °C. It was later revealed that the method of recording the temperature inside the vessel was flawed, resulting in a much higher temperature actually present within the still than that anticipated.

A sequence of events, which led to some confusion, followed the initial order to heat the sludge with steam. According to the inquiry, an operator, who had been called to remove the access hatch, arrived on site but went to lunch instead of carrying out the task.

The task of removing the sludge was under the control of the Area Manager, and he became concerned by the delay. However, two other men volunteered to open the hatch instead of waiting until the qualified operator returned from lunch.

An iron rake, which measured approximately 2.5 m in length, was used by two technicians to remove the sludge from the vessel. After an hour, it was decided that the rake needed to be extended to reach further into the still.

At about this time, the original operator arrived back from lunch. He noted that the still base inlet pipe had not been sealed and was immediately given the job of sealing the inlet feed.

The raking operation had been suspended temporarily until the extension was made to the rake. The manager in charge of the operation went to lunch and the technicians who had been carrying out the raking also left the vessel. Five minutes afterwards, a jet of flame erupted from the tank.

Such was the force of the jet of flame that a v-shaped scorch was made to the office building 55 m away. Windows were shattered and their frames melted. The blast caused a control panel, weighing more than 100 kg, to be blown onto its side.

The premise came under the Fire Certificates (Special Premises) Regulations 1976, which detailed three main escape routes from the office building. It was later found that one of the escape routes had breaches in the fire-resisting construction above a false ceiling. These breaches had resulted in smoke penetrating through the building and into the women's toilet, where a young girl was found overcome by smoke and toxic gases.

An extensive investigation was carried out by the Health and Safety Executive (HSE), which involved 270 staff days at a cost of more than £55 000. After thorough testing, the HSE proposed that the incident was probably caused by:

> Thermally unstable residues, in contact with steam, undergoing exothermic decomposition, producing enough energy to auto-ignite a flammable mixture of vapours or decomposition products.

Lessons Learned from the Incident

Many lessons were learned from this incident; the following is a summary of those lessons that were particularly important:

- Residues should be analysed, monitored and removed at regular intervals to prevent build-up of unstable impurities.
- Safe systems of work covering all aspects of operation and maintenance of all process plants should be established and defined in comprehensive instructions. This is extremely important where operations are undertaken at infrequent intervals.
- No assessment was made for the effects and potential hazards and risks of applying heat to the residues. As a result of the lack of risk assessment, the task was not properly planned.
- The design and location of buildings near chemical plant should be based on the potential for fire, explosion and/or toxic release and these potentials should be assessed by the process of risk assessment.
- Companies should regularly monitor, inspect and audit their own compliance with the performance standards in their fire certificates. The effect

of breaching fire-resisting structures should be thoroughly assessed by the process of risk assessment.

RISK ASSESSMENT

It is obvious from the above case that risk assessment plays an extremely important role in preventing fire and explosion, human suffering and damage to property. Fire risk assessment is not only for large chemical process plants or for complex premises. It is mandatory for employers to carry out fire risk assessment regardless of the size and complexity of the premises or work operation. However, it is fair to say that simple work premises require a much more simple risk assessment than that required by large industrial process plants.

The important aspect of fire risk assessment is that it matches the complexity of hazards and risks within a particular workplace. As a guide to the complexity of risk, the following can be used:

High-Risk Premises

A high-risk premise in relation to fire risk assessment might include a workplace where:

- there is a high life risk, e.g. large number of people employed or present;
- there are large quantities of flammable or highly flammable materials processed, stored or used; and
- there is a high risk of rapid fire spread through the premise, e.g. an older historical building that does not contain modern fire-resisting materials or a building with a large amount of false ceilings or other voids.

High-risk workplaces require comprehensive assessment of fire risks, which should be well detailed and should be communicated to all persons concerned. Regular review of the risk assessments should be made, taking into consideration any changes in structure, work operations or technology available to improve fire resistance of materials.

Medium-Risk Premises

A medium-risk premise in relation to fire risk assessment might include a workplace where:

- there are a large number of people present; however, the building may be modern in construction and may contain a fire for a sufficient period of time to allow all occupants to escape;
- there are combustible materials present, but these are stored and used in areas with good fire-resisting material; and
- rapid fire spread is unlikely owing to modern construction methods.

Once again, fire risk assessment is important, and should cover work activities and the people who are likely to be involved. However, a less comprehensive assessment is needed, and the assessment can be reviewed less frequently than the assessments for high-risk premises.

Low-Risk Premises

A low-risk premise in relation to fire risk assessment might include a workplace where:

- a small number of people are employed or present in the workplace;
- very minimal amount of combustible materials are used and stored and even in places where these are used or stored, they are kept in properly designated areas; and
- rapid fire spread is unlikely owing to the size and/or construction of the premise.

A fire risk assessment for such premises may be quite simple, and in such premises, only significant risks need to be recorded. It is important that, although the premise may be of low risk, complacency of management and employees does not put people at unnecessary risk. Fires can still occur and can kill and injure people even in the smallest of workplaces.

Regardless of the category of risk, fire risk assessments should be carried out for all workplaces, even if the workplace has been subject to previous approvals by the various enforcing authorities for other fire safety, licensing or building legislation.

It is important to remember that any changes to fire precautions proposed as a result of carrying out a fire risk assessment do not conflict with the controls imposed by other legislation. If any doubt exists, the local fire authority should be contacted before any changes, especially structural changes or changes involving the means of escape, are made.

Another important point is where other employers share premises. After carrying out fire risk assessments, all employers sharing a premise or building must be informed of the results of the assessment and of any control measures that are to be put in place to control the risk of fire.

THE FIRE RISK ASSESSMENT PROCESS

Fire risk assessment can be divided into the following stages:

Stage 1 Identify fire hazards.
Stage 2 Identify the location of people (and the types of people) at significant risk in the event of fire.
Stage 3 Evaluate the risks.
Stage 4 Consider existing controls that are in place.

Stage 5 Consider further controls that may need to be put in place.
Stage 6 Consider other legislation that may be applicable.
Stage 7 Record the findings.
Stage 8 Produce an action plan and implement the plan.
Stage 9 Communicate the findings to employees and others.
Stage 10 Review the assessment.

Stage 1 Identify Fire Hazards

A hazard can be defined as 'something with the potential to cause injury to people, damage to property or impact on the environment'.

Life in general is full of hazards. From getting up in the morning to returning to bed at the end of the day, people are continually faced with hazards. Crossing the road, driving a car or using public transport, lifting equipment at work, using machinery and the use of chemicals and other substances are some of the hazards that are encountered daily by employees.

Fire itself is a major hazard. A fire in a large building certainly has the potential to cause injury or even death to people. It has the potential to cause damage to property – many fires cause millions of pounds of damage – and a fire has a potential impact on the environment e.g. smoke damage and fire-fighting water damage.

In terms of fire, there are many hazards in the workplace. It is often simpler to categorise hazards into three main areas. For a fire to start and to continue burning, it requires three things. These are:

- Oxygen
- Heat or ignition source
- Fuel.

Oxygen

Oxygen is usually present in the air that we breathe but it may be enriched in certain circumstances.

The main source of oxygen for a fire in the workplace comes either naturally through doors and windows or can be forced into the building by ventilation equipment. The latter may be capable of both extracting and introducing air.

In addition to natural and mechanical systems of introducing oxygen for a fire, there are other sources. These include the following:

- Chemicals such as oxidizing agents that can provide a source of oxygen for a fire, which may result in a rapid spread of fire through a premise. These chemicals are dealt with in a later chapter.
- Oxygen supplies from cylinders and piped systems can provide a source for a fire to spread quickly. These include oxygen for medical purposes and oxygen for welding or cutting purposes.

Sources of Heat

There are many potential sources of heat in the workplace and an important part of the risk assessment process is to try and identify these sources. It may be that sources of heat exist but are adequately controlled; however, the first stage of the risk assessment process is to identify them regardless of how they are presently controlled.

Sources of heat or ignition can include the following:

- Smoking materials
- Electrical equipment
- Gas or oil-fired heating equipment
- Cooking equipment
- Naked flames
- Vehicle engines
- Other engines, e.g. emergency generators
- Static electricity
- Friction, e.g. loose bearings or drive belts
- Flammable dust build-up on surfaces or equipment
- Machinery
- Lighting equipment
- Hot processes, e.g. welding, cutting or grinding
- Arson
- Hot surfaces.

Sources of Fuel

Anything that is combustible is a potential source of fuel. The most common sources of fuel include the following:

- Flammable solids, e.g. paper, wood, plastics, waste material
- Flammable liquids and solvents, e.g. petrol, paraffin
- Flammable gases, e.g. propane, butane, and acetylene
- Furnishings such as foam used in upholstered furniture
- Flammable dusts, e.g. flour, aluminium.

In summary, the first stage of the risk assessment process is to identify hazards: potential sources of oxygen; potential sources of ignition; and potential sources of fuel. The incident and resulting explosion and fire at Hickson and Welch illustrated how the failure to properly identify hazards arising from work activities can have such a tragic outcome. Had suitable and sufficient risk assessments been carried out on the task of removing sludge and the hazards been identified properly, suitable control measures could have been put in place, which may have controlled the risk.

A comprehensive checklist is provided at the end of this chapter, which can be used as a basis for identifying hazards found in the workplace. The checklist should be used systematically to gather information regarding the hazards that may be present. Once hazards have been identified, the next part of the process is to identify how many people are likely to be affected by those hazards.

Stage 2 Identify the Location of People at Significant Risk in the Event of Fire

The main priority when dealing with fire precautions is to ensure that people can escape safely in the event of a fire. A workplace must have adequate means of detecting a fire and giving warning to all occupants that a fire has been detected. In addition, workplaces must have adequate means of escape to ensure that, in the event of a fire, occupants can escape safely through protected routes to a place of safety. Escape routes have to be protected from rapid smoke and heat spread.

When considering people who are likely to be affected by fire, it is important to consider not only employees but also other categories of people who may be present. The following questions should be asked as part of the fire risk assessment process:

- How many people normally occupy the workplace?
- Are contractors employed, and if so, are they employed frequently or only occasionally?
- What do contractors do in the event of a fire alarm sounding?
- Are disabled persons likely to be present in the workplace and, if so, how will they be able to hear or see fire warning systems?
- If disabled persons in wheelchairs are part of the workforce or are likely to be present in the workplace, have provisions been made to assist them in the event of a fire alarm sounding?
- Do people know what to do in the event of a fire alarm sounding? Do they know, without any doubt, what their actions should be?
- Has a fire warden system been set up to assist with evacuation?
- Do people know what to do after leaving the building in the event of a fire?
- Have muster points been established outside the building or the premises?
- Is there a visitor's register?
- How will management ensure that all persons have been evacuated in the event of a fire?

By answering the above questions, a more detailed picture can be developed regarding the number and type of persons who are likely to be present within the workplace and how these people will respond in the event of a fire. If some of the answers to the above questions cannot be found, these areas need to be explored more fully during the next part of the risk assessment process.

A major aspect of fire is how people will react in the event of a fire alarm sounding or, even worse, in an actual fire situation. Human behaviour is a complex subject and is an extremely important part of the risk assessment process. Human behaviour in fire is dealt with in a separate chapter.

Stage 3 Evaluation of the Risk

Once hazards have been identified and the types of people who may be affected by fire have been identified, the next stage of the process is to evaluate the actual risk from the hazards present.

Evaluation of risk should consider what is already in place and whether the existing measures are adequate. Generally, the following must be considered:

- What means are in place to detect an outbreak of fire.
- What provision has been made to warn everyone in the premise or building that an outbreak has been detected.
- What arrangements have been made to ensure the means of escape is adequate.
- What provision has been made for fire-fighting equipment.
- Have people been given training in the action to take should a fire occur.

Perhaps some or all of the above provisions already exist. Perhaps none of the above are in place. Stage three of the risk assessment process evaluates what is in place and what further action may have to be put in place to adequately control the risk from fire.

Definition of Risk

It has previously been stated that a hazard can be defined as, 'something with the potential to cause injury to people, damage to property or impact on the environment'. Many things in life present hazards. However, do they also present a risk?

Risk can be defined as, 'the combination of the likelihood and the consequences of a hazard actually occurring'. Risk therefore has two elements:

- the likelihood that the hazard will occur, and;
- the consequences of the hazard if it does occur.

For example, suppose two employees use acetylene equipment in the workplace. Suppose an initial hazard-spotting exercise, using the checklist at the end of this chapter, has revealed the following:

- Both operators have been trained in the use of the equipment.
- Both operators have received regular refresher training in the use of the equipment.
- The equipment is always stored in an external store overnight.
- A competent gas supplier checks the equipment on a regular basis.

However, the hazard-identification exercise had also revealed that, although the operators remove any combustible debris from the area of welding, no attempt is ever made to ensure that a fire extinguisher is readily available during work with the acetylene equipment.

From the above example, it is clear that a fire hazard exists – that of using acetylene equipment. However, is there a great risk? Well, probably not. Both operators are trained and competent, the equipment is stored properly when not in use and the equipment is checked regularly by a qualified gas supplier. With those control measures in place, the risk from a very high hazard has been reduced to a low level. Provided all the above precautions continue to be put in place, the risk of a fire involving the acetylene equipment is negligible.

Reasonably Practicable

The Management of Health and Safety at Work Regulations 1999 requires employers to put in place 'reasonably practicable' measures to control risk. The employer in the above example has indeed put in place many control measures to control the risk of fire. However, it may be argued that it would also be reasonably practicable for the operators to ensure that an additional fire extinguisher is nearby when using the equipment. This final control measure would ensure that all reasonable precautions have been taken to prevent, or deal with, a fire while using acetylene equipment.

The above simple example illustrates the concept of the evaluation of risk. Once hazards have been identified, the employer must look carefully at what is in place already and decide whether they are reasonable for the circumstances. If not, additional controls must be put in place.

A difficult question to answer is often, 'what is reasonable?'. It is possible to reduce most risks in the workplace; however, to do so would require large amounts of money. Using the above-described acetylene equipment as an example, it would be possible to build a fire enclosure around the area where the acetylene is used. The structure could be made from materials that would maintain its integrity under fire conditions for three hours. The enclosure could be fitted with sprinklers and automatic fire detection. By putting in place these types of controls, the risk from fire is greatly reduced. However, all these controls would be very costly in terms of money. In addition, it probably would not be practical for all welding to be done in one area and so the enclosure would have to be portable, which is not too practical! Therefore, it is not considered reasonable to install such elaborate control measures.

This concept of identifying hazards and evaluating risk from those hazards and then putting in place reasonable controls is the basis of modern health and safety legislation.

Using the above example, it might however be considered reasonable to ensure that additional fire extinguishers are provided when work with acetylene is undertaken. This additional measure would not be too expensive. In fact, it probably would not involve any additional costs.

A Risk-Evaluation Model

There are many models of risk evaluation, from the very complex involving mathematical formulae to the very simple using only brief numerical scores. If fire risk assessment is to be carried out on processes involving highly flammable materials or explosion hazards, a complex model of risk evaluation

Table 2.1 Risk-level estimator

	Unlikely (1)	Likely (2)	Very likely (3)
Minor (1)	1	2	3
Major (2)	2	4	6
Serious (3)	3	6	9

is perhaps appropriate. In most circumstances, and indeed in most workplaces, a simple numerical model would suffice.

One model of risk evaluation that may be helpful is provided above. The model uses a simple risk-level estimator, which helps to calculate the risk from a particular hazard. From there, a risk-based control plan can be developed according to the level of risk calculated.

The risk-level estimator (Table 2.1) uses a combination of probability and consequence to calculate the outcome in terms of risk.

Probability

Unlikely (1) – the task that involves the hazard is not normally carried out or is carried out only occasionally, perhaps once per month.

Likely (2) – the task that involves the hazard is carried out frequently, perhaps weekly.

Very likely (3) – the task that involves the hazard is carried out routinely, perhaps daily or more than once per day.

Consequence

Minor (1) – a fire may be possible from the hazard that exists; however, owing to the present control measures, the escape routes and the number of people in the area, it is unlikely that a person may become trapped or overcome by heat and smoke.

Major (2) – A fire may be possible from the hazard that exists and the fire may develop quickly, resulting in a higher probability that a person may become trapped. In addition, given the reduced level of control at present or the large numbers of people in the area, the probability that people may be overcome by heat and smoke is higher than the above category.

Serious (3) – A fire is likely to start from the hazard and, given the reduced level of controls at present, would probably trap a person or impede their ability to escape. The hazard is such that a fire may develop rapidly. In addition, large numbers of people are present in the area, which significantly increases the probability of a person being overcome by heat and smoke.

Using the above risk-level indicator, a risk rating of each fire hazard can be obtained. This risk rating can be used to help determine whether or not adequate controls are in place.

Stage 4 Consider Existing Controls

Whenever the process of risk assessment is carried out, it is important to consider existing control measures. Risk calculations should be made after considering existing controls.

Existing controls may include the following:

- Means of escape and its construction.
- Means for evacuating the building and taking a roll call.
- Any fire-fighting equipment that is currently provided and the provision of maintenance to ensure those items are kept in working order.
- Any procedures that exist for evacuation and for fire-fighting arrangements
- The system for notifying the fire brigade that a fire is suspected.
- Any training that has been provided for employees, e.g. fire drills, fire-fighting courses.
- Procedures for ensuring that equipment is maintained in correct working order, e.g. electrical equipment and machinery.

Stage 5 Further Controls

Once an assessment of the current risk, which should take into consideration existing controls, has been made, a decision has to be made regarding the acceptability of the current risk. In other words, is the current level of risk reasonably practicable or should other control measures be put in place.

If an employer measures risk and is satisfied that it is very low, with little probability of a person being harmed, then any further action that would be required will be very minimal. However, on the other hand, if a very high level of risk exists, even with many control measures in place, the employer must consider further controls to reduce the risk to a more acceptable level.

Risk-Based Action Plan

A risk-based action plan is provided in Table 2.2 to help make decisions about further control measures. Although this can be used for guidance, it is not intended as definitive. Many factors need to be assessed for further action to be planned and implemented.

Stage 6 Consider Other Legislation that may be Applicable

The risk assessment process is a very useful tool in identifying other legislation that is applicable. For example, a fire risk assessment may identify that signs need to be displayed to identify exits on escape routes. All safety signs should comply with the Health and Safety (Safety Signs and Signals) Regulations 1996, which require safety signs to be of a certain colour and design.

Some other regulations that may be applicable to fire risk assessment are as follows.

Table 2.2 Risk-based action plan

Risk rating	Action and timescale
1	The current level of risk is considered acceptable and therefore no further action is required. Existing controls should be maintained in their present condition.
2	Although the current level of risk is acceptable, consideration should be given to a cost-effective method of reducing the risk to a lower level. Monitoring is required to ensure that the existing controls are maintained.
3–4	A moderate fire risk exists, and efforts should be made to reduce the risk. The cost of control measures should be carefully measured and selected to give a cost-effective return. Risk-reduction measures should be implemented within a defined time period. Monitoring is required to ensure that the existing controls, and additional controls, are maintained.
6	Work should not be started until the fire risk has been reduced. Considerable resources may have to be allocated to reduce the risk. Where the risk involves work in progress, immediate action should be taken. Monitoring is required to ensure that the existing controls, and additional controls, are maintained.
9	Work should not be started or continued until the fire risk has been reduced. If it is not possible to reduce risk even with unlimited resources, work has to remain prohibited.

Provision and Use of Work Equipment Regulations 1998

These regulations cover work equipment including machinery and require the employer to ensure that all items of work equipment are properly installed, used and maintained. Items of electrical equipment and machinery may present a high fire risk and it is essential that they be maintained in accordance with the manufacture's instructions.

Workplace (Health, Safety and Welfare) Regulations 1992

These regulations require that the physical condition of the workplace be such that it does not present a hazard to employees. This includes adequate lighting, ventilation, trip and slip hazards and good housekeeping. The latter point is particularly important as far as fire risk is concerned. Many fires are started in the workplace through poor standards of housekeeping.

Electricity at Work Regulations 1989

These regulations relate to the design, construction and maintenance of electrical systems and work activities on or near electrical installations. Since electrical installations present a high level of fire risk, compliance with these regulations will help to reduce the risk of fire in the workplace.

Control of Major Accident Hazards Regulations 1999

These regulations relate to the storage and use of dangerous substances and require an employer to submit a safety report, detailing the type of materials and their quantities stored or used at the premises. In addition, an on-site emergency plan should be developed that details the actions necessary to warn that a major incident has occurred and the actions necessary to minimize the consequences of such an incident. An off-site emergency plan is also required to warn the local community that an incident has taken place and the action they have to take in such an event.

Highly Flammable Liquids and Liquefied Petroleum Gases Regulations 1972

These regulations are applicable to the storage and use of highly flammable liquids and gases. The regulations require adequate storage facilities in the workplace to be provided for the storage of the above materials, which includes suitable fire-resisting structures, adequate marking of the storage facilities and proper procedures for the use of the materials.

Gas Safety (Installation and Use) Regulations 1998

These regulations require employers to ensure that competent persons install and maintain gas systems that are used in the workplace.

The above is not a comprehensive list of applicable regulations; however, they provide an indication of some of the regulations that are important as part of the fire risk assessment process.

Stage 7 Record the Findings of the Assessment

If there are five or more employees, the results of the fire risk assessment must be recorded. There are many different methods of recording and storing information relating to risk assessments. However, it is important to choose a method that is:

- easy to use
- easy to communicate the findings to relevant people, e.g. employees, contractors and other persons who may share the premises
- easy to provide to the enforcing authority
- easy to file and retrieve if necessary in the future.

A suggested format is provided below, which can be used in conjunction with the checklist mentioned previously and will provide a simple system for recording the findings of a fire risk assessment.

Assessment number					
Assessor		Assessment date		Location assessed	
Item number	Hazard	People at risk	Existing controls	Risk	Further controls
Other applicable regulations				Details	

Stage 8 Produce an Action Plan

After the risk assessment has been carried out and the findings recorded in a suitable format, the next stage of the process is to develop an action plan. It is essential that the findings of the assessment be acted upon. In fact, when the enforcing authority inspect the premises in accordance with the Fire Precautions (Workplace) Regulations, they will require to see what action has been taken and what remains to be done.

An important aspect of the action plan is to include appropriate time periods for the action points. These will very much depend upon the risk that exists at present. For example, if there is a high risk of a fire starting from a particular hazard, action might have to be taken very quickly. However, if a low risk exists, requiring minimal additional controls to be put in place, the time period for this might be a little longer.

Once again, a simple form, such as the following, can be used to record such details.

Assessment number				
Assessor		Assessment date		Location assessed
Item number	Action	Responsibility	Target date	Completion date

Stage 9 Inform Others of the Risk

The next stage of the fire risk assessment process is to inform other people of the risks from fire. This will include employees, contractors and possibly visitors, depending on the nature of risk.

The higher the risk of fire, the more that will have to be done as far as informing employees is concerned. Information should be provided in the most suitable form. If an employee is involved in a high-risk activity, that employee might need significant training in the risks and preventative measures to take to minimize fire. However, an office worker may only need periodic training in the correct procedure to be followed in the event of fire.

Information can be provided in whatever form is considered necessary whether it be formal training in a classroom environment, by written procedure or by on-the-job training. A combination of all three methods often works best.

In addition to providing information to employees and contractors, there are other circumstances where further training may be needed. A special procedure may have to be developed for disabled persons and this needs to be communicated, perhaps on an individual basis. Also, where young persons are employed, their parents or guardians must be provided with information regarding the risks associated with their work activities in the workplace.

Stage 10 Review the Risk Assessment

Sooner or later, changes that have an effect on the fire risk assessment and the control measures that have been put in place may be introduced into the workplace. Changes might include the following:

- A change in the number of employees.
- Additional disabled persons working in the premise.
- A change in the work process.
- New furniture or fittings.
- New plant, machinery or other equipment.
- A new building or an extension to the existing premises.
- New substances or a change to the existing stock of chemicals and substances.

Every time an item from the above list is changed, it has the potential to introduce new hazards, which might not have been taken into consideration previously. It is, therefore, essential that a reassessment be carried out whenever such changes are made. This will ensure that the risk assessments and the control measures put in place are always current. It is also important to reassess at other intervals, such as:

- when a fire occurs, regardless of its size or nature;
- when a near-miss incident occurs, which might have started a fire, or;
- after a specified time period.

Reassessment After A Specified Time Period

- Low risk premises – reassessment should be carried out every two years.
- Medium risk premises – reassessment should be carried out every 18 months.
- High risk premises – reassessment should be carried out every 6–12 months.

Competent Persons

A competent person should carry out fire risk assessments. It is the employer's responsibility to appoint someone who has both the experience and the qualifications to carry out such a task. In addition, the person must be competent in devising and applying suitable control measures. On deciding whom to appoint, employers themselves need to know and understand the work involved, the principles of risk assessment and current legislation relating to fire risks in the workplace.

SUMMARY

Risk assessment is a powerful tool as far as making the work environment a safer place is concerned. It is a concept used in modern health and safety legislation, which centres on the principle of being reasonably practicable. Employers must ensure that they have taken reasonable steps to reduce risk to a level as low as reasonably practicable.

Through the process of risk assessment, an evaluation can be made as to the current level of risk from fire in the workplace. From there, decisions should be made on whether additional controls have to be put in place.

REFERENCES

HSE Books (1994) *The Fire at Hickson and Welch Ltd*, HMSO.
HSE Books (1999) *Fire Safety–an employers guide*, HMSO.

APPENDIX FIRE RISK ASSESSMENT CHECKLIST

Location:
Assessment Date:
Assessor:

1 General

Item	Potential hazard	Details	Further assessment
a.	Provide a general description of the building.		
b.	How many occupants are usually present in the building?		
c.	Are mobility-disabled persons present in the building?		
d.	Do persons suffering from reduced hearing, visual impairment or mental impairment occupy the building?		
e.	Is the building occupied on a 24-hour basis?		
f.	Are security guards employed to patrol the building?		

2 Structural Features

Item	Potential hazard	Details	Further assessment
a.	Does the building contain compartmentation?		
b.	Have openings been created in fire-separating walls, e.g. for pipes, cables, ducting?		
c.	Is there a mechanical ventilation or air-conditioning system in the building?		
d.	Is the building fitted with one or more passenger lifts?		
e.	Is the building fitted with one or more goods lifts?		
f.	Does the building contain areas where false ceilings or false walls have been installed?		
g.	Is the building clad with combustible materials?		

3 Sources of Ignition

Item	Potential hazard	Details	Further assessment
a.	Is there a no-smoking policy for employees, contractors and other persons?		
b.	Are portable items of electrical equipment used in the building?		
c.	Are gas or oil-fired heating appliances used in the building?		
d.	Is there a cooking facility within the building?		
e.	Are naked flames present in the building?		
f.	Do vehicles, including fork-lift trucks, operate within or outside the building (other than private vehicles)?		
g.	Are emergency generators used as a secondary source of power to the building?		
h.	Is it possible to generate static electricity by work activities within the building (other than static electricity produced by lightning)?		
i.	Have lightning conductors been fitted externally to the building?		
j.	Are there items of machinery or equipment used that contain bearings or drive belts?		
k.	Are there any processes present that are likely to generate flammable dust?		
l.	Are there items of machinery present in the building?		
m.	Is portable lighting equipment used within the building, e.g. halogen lamps?		
n.	Are there items of equipment that are likely to produce hot exposed surfaces, e.g. uninsulated steam pipes, hot machinery surfaces?		
o.	Is hot work carried out within or near to the building, e.g. welding, cutting, and grinding?		
p.	Have there been cases of arson within the building or within the grounds of the building?		
q.	Have there been reported cases of arson within the neighbourhood?		

4 Combustible Materials

Item	Potential hazard	Details	Further assessment
a.	Is the quantity of combustible materials being stored or used kept to a minimum?		
b.	Are the combustible materials kept clear of fire escape routes?		
c.	Are the combustible materials kept away from heat sources?		
d.	Are highly flammable materials stored within fire-resistant stores?		
e.	Are the quantities of highly flammable materials that are stored and used kept to a minimum?		
f.	Is combustible waste collected and removed from the premises at frequent intervals?		

5 Means of Escape

Item	Potential hazard	Details	Further assessment
a.	How long will it take for all occupants to escape to a place of safety once a fire has been detected?		
b.	Are there enough fire exits?		
c.	Are the types of exits suitable for the number of people likely to use them?		
d.	Are all fire exits easily opened?		
e.	Have appropriate disabled refuge areas been provided?		
f.	Is the integrity of fire doors and associated escape routes checked on a regular basis?		
g.	In the event of fire, could all exits be affected or will at least one route remain available?		
h.	Do exits lead to a place of safety?		
i.	Are all escape routes easily identifiable?		
j.	Are all escape routes clear from obstruction?		
k.	Are all escape routes adequately illuminated?		
l.	Have personnel been trained in the action to follow in the event of fire?		
m.	Are there instructions about the means of escape?		
n.	Are external escape routes properly maintained?		

6 Fire Warning and Detection

Item	Potential hazard	Details	Further assessment
a.	Can the existing detection system discover a fire quickly enough to raise the alarm in time for all occupants to escape to a place of safety?		
b.	Can the means for giving warning be clearly heard throughout the premises?		
c.	Is there a back-up power supply for the fire warning system?		
d.	Do employees know how to operate the fire warning system?		
e.	Are there instructions for employees on how to operate the fire warning system?		
f.	Is the fire warning and detection system tested in accordance with appropriate British Standards and/or manufacturers instructions?		

7 Means of Fire-Fighting

Item	Potential hazard	Details	Further assessment
a.	Are fire extinguishers suitable for the purpose?		
b.	Are fire extinguishers of suitable capacity?		
c.	Are there enough extinguishers throughout the workplace?		
d.	Are extinguishers located in the correct positions throughout the workplace?		
e.	Can users gain access to fire extinguishers without exposing themselves to risk?		
f.	Are the locations of fire extinguishers obvious?		
g.	Have people been trained in the use of fire extinguishers?		
h.	Are fire extinguishers tested in accordance with the manufacturer's instructions?		

8 Primary and Emergency Lighting

Item	Potential hazard	Details	Further assessment
a.	Is the primary lighting system adequate and functioning properly?		
b.	Is the emergency lighting system tested in accordance with appropriate British Standards and/or the manufacturer's instructions?		

9 Fire Evacuation

Item	Potential hazard	Details	Further assessment
a.	Are there written procedures in place for evacuation in the event of fire?		
b.	Have all personnel been trained in the procedures?		
c.	Is there a system of training for contractors in the evacuation procedures?		
d.	Is there a visitor's register?		

10 Fire Limitation

Item	Potential hazard	Details	Further assessment
a.	Is compartmentation provided within the premises?		
b.	Is there any form of structural fire resistance within the premises?		
c.	Is there a smoke ventilation system?		
d.	Is there a pressurization system?		

11 Fire Notices

Item	Potential hazard	Details	Further assessment
a.	Are notices provided on fire doors?		

12 Fire Safety Checks

Item	Potential hazard	Details	Further assessment
a.	Are checks carried out to ensure that windows and doors are closed when the premises are not occupied?		
b.	Are checks carried out to ensure that electrical equipment is turned off and unplugged when the premises are not occupied?		
c.	Are checks carried out to ensure that smoker's materials are not left smouldering?		
d.	Are checks carried out to ensure that naked flames are extinguished when the premises are not occupied?		
e.	Are checks carried out to ensure that the premises are left secure when not occupied?		

13 Equipment and Plant

Item	Potential hazard	Details	Further assessment
a.	Are all items of electrical equipment, plant and machinery tested and maintained in accordance with the manufacturer's instructions?		

14 Electrical and Gas Installations

Item	Potential hazard	Details	Further assessment
a.	Has the electrical system within the premises been tested in accordance with the Electricity at Work Regulations?		
b.	Has the gas system within the premises been tested in accordance with the Gas Safety (Installations and use) Regulations?		

15 Liaison

Item	Potential hazard	Details	Further assessment
a.	Is liaison with the local fire authority maintained?		
b.	Is there adequate access for fire appliances?		
c.	Are external assembly points adequately marked?		
d.	Are fire hydrants and sprinkler valves adequately marked?		

16 Emergency Plan

Item	Potential hazard	Details	Further assessment
a.	Is there an emergency plan?		

17 Logbooks

Item	Potential hazard	Details	Further assessment
a.	Is there a logbook to record tests carried out on fire alarm and fire detection equipment?		
b.	Is there a logbook to record tests carried out on fire extinguishers and other fire-fighting equipment?		
c.	Is there a logbook to record tests carried out on emergency lighting equipment?		
d.	Is there a logbook to record fire training for employees?		
e.	Is there a logbook to record tests on fire doors and associated equipment?		

18 Legislation

Item	Potential hazard	Details	Further assessment
a.	What other legislation might be relevant to prevent fire within the premises?		

19 Review of Risk Assessment

Item	Potential hazard	Details	Further assessment
a.	Have arrangements been made to review the fire risk assessment at regular intervals?		

3 Case Study

This case study, which is based on a real company, is provided to help in gaining an understanding of the practical application of the risk assessment process. A brief outline of the company, its employees and the premises where the company is located has been provided. A risk assessment checklist (Appendix A) and an actual assessment (Appendix B) have been completed. The findings of the risk assessment have resulted in an action plan (Appendix C) being produced.

The process of completing a checklist, carrying out an assessment and finally producing an action plan should help to apply the principles already discussed in previous chapters. The information that follows is the type of information that the enforcing authorities are likely to require.

XYZ ENGINEERING LTD

XYZ Engineering Ltd is a light engineering company, which specializes in the repair of hydraulic and electrical pumps and motors. The premises occupied by XYZ is part of a large industrial unit. XZY occupies approximately 70% of the premises, while the other 30% is occupied by a ceramic manufacturing company, which employs 12 people. XYZ Engineering employs 55 employees in the following departments:

- 10 employed within the general office, and;
- 45 employed within the workshop.

One employee in the office is permanently confined to a wheelchair. In addition to employees, contractors are employed to carry out specialized work within the workshop. At any one time, there may be as many as five contractors in the workshop. Tasks carried out within the workshop include the following:

- Welding and grinding operations.
- An electrical repair area.
- A dispatch area where equipment is loaded onto vehicles.
- A small paint-application area.

Some of the important features of the premises are as follows:

- Evacuation time from the premises would usually take about two minutes.
- Two exits are available from the office and four from the workshop; however, one of the exits is seldom used.
- All exits open to a place of safety.
- Emergency exit signs have been provided throughout the premises.
- Break glass points have been provided near final exit doors.

- Various fire extinguishers have been provided throughout the premises.
- A company has been contracted to service the fire alarm and fire-fighting equipment.

Some other important information that would be essential for risk assessment purposes includes the following:

- A local exhaust ventilation system has been provided over one of the areas used for soldering.
- Various items of equipment are stored above the dispatch area.
- Various items of portable electrical equipment are used.
- Lorries and fork-lift trucks are usually present within the dispatch area.
- Paints and thinners are used and stored within the premises.
- Several cases of arson have been reported within the neighbourhood.

The premises were built in 1980 and consist of brick construction with a tiled roof. There are no particular problems with the construction as far as fire spread is concerned. Although both companies share the same premises, walls made from fire-resisting construction separate each unit.

The premises were last inspected by the local fire authority 18 months ago and a fire certificate was issued under the Fire Precautions Act 1971. However, three months ago the local fire authority asked each of the companies that share the premises to submit a fire risk assessment.

No formal meetings take place between the occupiers of the two companies and no means of sharing information about the risks in each company is evident.

APPENDIX A FIRE RISK ASSESSMENT CHECKLIST

Location: XYZ Engineering Ltd
Assessment Date: August 2001
Assessor: John Smith, General Manager

1 General

Item	Potential hazard	Details	Further assessment
a.	Provide a general description of the building.	The premises, which is part of a larger building, is divided into an office and a workshop area. Within the workshop, four areas are located: welding and grinding; electrical repair; dispatch; and paint application. A fire certificate has been provided by the local fire authority, certificate number B/120/002.	No.
b.	How many occupants are usually present in the building?	55 employees in the following locations: office, 10; workshop, 45. In addition, there are often contractors who carry out specialized work within the workshop. At any one time, there may be a maximum of five contractors at work. Also, visitors may be present in both the office and workshop areas.	No.
c.	Are mobility-disabled persons present in the building?	One employee in the office area is permanently confined to a wheelchair. No special provision has been made for evacuating this employee in the event of fire.	Yes. Assess whether the existing evacuation method is suitable and assess the individual needs of the disabled person.

(*Continued*)

Item	Potential hazard	Details	Further assessment
d.	Do persons suffering from reduced hearing, visual impairment or mental impairment occupy the building?	No.	No.
e.	Is the building occupied on a 24-hour basis?	No.	No.
f.	Are security guards employed to patrol the building?	No.	Yes. Assess security, particularly because arson is a problem in the area.

2 Structural Features

Item	Potential hazard	Details	Further assessment
a.	Does the building contain compartmentation?	No. Not required by building regulations.	No.
b.	Have openings been created in fire-separating walls, e.g. for pipes, cables, ducting?	No.	No.
c.	Is there a mechanical ventilation or air-conditioning system in the building?	No. However, there is a local exhaust ventilation system over one of the soldering units in the electrical workshop area.	Yes. Assess fire risk in ventilation system.
d.	Is the building fitted with one or more passenger lifts?	No.	No.
e.	Is the building fitted with one or more goods lifts?	No.	No.
f.	Does the building contain areas where false ceilings or false walls have been installed?	Generally, no. However, there is a floored area above dispatch, which is used to store various items of electrical stock, cleaning material and office supplies.	Yes. Assess the possibility of fire starting from the storage area above vehicles.
g.	Is the building clad with combustible materials?	No.	No.

3 Sources of Ignition

Item	Potential hazard	Details	Further assessment
a.	Is there a no-smoking policy for employees, contractors and other persons?	No.	Yes. Assess the possibility of fire starting from smoker's materials.
b.	Are portable items of electrical equipment used in the building?	Yes. Various items of portable electrical equipment are used.	Yes. Assess portable equipment.
c.	Are gas or oil-fired heating appliances used in the building?	No.	No.
d.	Is there a cooking facility within the building?	No.	No.
e.	Are naked flames present in the building?	Yes, used to light welding equipment.	Yes. Assess possible ignition of combustible materials.
f.	Do vehicles, including fork-lift trucks, operate within or outside the building (other than private vehicles)?	Yes. These include lorries and a fork-lift truck.	Yes. Assess possibility of engine fire or vehicles starting a fire from combustible material.
g.	Are emergency generators used as a secondary source of power to the building?	No.	No.
h.	Is it possible to generate static electricity by work activities within the building (other than static electricity produced by lightning)?	Not likely.	No.
i.	Have lightning conductors been fitted externally to the building?	Yes.	No.
j.	Are there items of machinery or equipment used that contain bearings or drive belts?	No.	No.
k.	Are there any processes present that are likely to generate flammable dust?	No.	No.

(*Continued*)

Item	Potential hazard	Details	Further assessment
l.	Are there items of machinery present in the building?	No. Only portable tools and bench grinders.	Yes. Assess possibility of fire from bench equipment.
m.	Is portable lighting equipment used within the building, e.g. halogen lamps?	No.	No.
n.	Are there items of equipment that are likely to produce hot exposed surfaces, e.g. uninsulated steam pipes, hot machinery surfaces?	No.	No.
o.	Is hot work carried out within or near the building, e.g. welding, cutting and grinding?	Yes, welding and grinding.	Yes. Assess possibility of ignition of combustible materials. See 3.e.
p.	Have there been cases of arson within the building or within the grounds of the building?	Yes. Several cases of arson have been reported, mainly involving skips within the grounds.	Yes. Assess possibility of deliberate ignition outside the premises.
q.	Have there been reported cases of arson within the neighbourhood?	Yes, as above.	Yes, as above.

4 Combustible Materials

Item	Potential hazard	Details	Further assessment
a.	Is the quantity of combustible materials being stored or used kept to a minimum?	Yes. The quantity of paints and thinners is very small and is kept to a minimum. These are stored in proper metal storage units.	No.
b.	Are the combustible materials kept clear of fire escape routes?	No. One of the escape routes from the workshop usually has a quantity of packing materials stored nearby.	Yes. Assess current storage of waste materials.

(Continued)

Item	Potential hazard	Details	Further assessment
c.	Are the combustible materials kept away from heat sources?	Yes.	No.
d.	Are highly flammable materials stored within fire-resistant stores?	Yes. These are stored in suitably marked units and the quantities are kept to a minimum.	No.
e.	Are the quantities of highly flammable materials that are stored and used kept to a minimum?	Yes.	No.
f.	Is combustible waste collected and removed from the premises at frequent intervals?	No. A quantity of waste packaging is usually stored near one of the escape routes.	Yes. Assess waste storage methods.

5 Means of Escape

Item	Potential hazard	Details	Further assessment
a.	How long will it take for all occupants to escape to a place of safety once a fire has been detected?	Under 2 minutes.	No.
b.	Are there enough fire exits?	Yes. Office area–two exits, one into the yard and one into the workshop. Workshop–four exits, one into the office area, one through dispatch, one at the other end of the workshop, and one in the middle of the workshop, which leads into the yard. All exits are positioned to comply with guidance on travel distances, which means that occupants can escape to a place of safety within a reasonable time.	No.

(*Continued*)

Item	Potential hazard	Details	Further assessment
c.	Are the types of exits suitable for the number of people likely to use them?	Yes.	No.
d.	Are all fire exits easily opened?	No. One exit at the middle of the workshop is never usually used and was found to be tight to open.	Yes. Assess possible problem with means of escape.
e.	Have appropriate disabled refuge areas been provided?	Not required.	No.
f.	Is the integrity of fire doors and associated escape routes checked on a regular basis?	No integrity checks are carried out.	Yes. Assess possible problem with fire doors.
g.	In the event of fire, could all exits be affected or will at least one route remain available?	At least one route would remain available.	No.
h.	Do exits lead to a place of safety?	Yes, into the yard or the car park.	No.
i.	Are all escape routes easily identifiable?	Yes. All exits have illuminated signs above them.	No.
j.	Are all escape routes clear from obstruction?	No. Waste material is often stored near one of the exits in the workshop.	See 4.f.
k.	Are all escape routes adequately illuminated?	Yes.	No.
l.	Have personnel been trained in the action to follow in the event of fire?	No training given to employees.	Yes. Assess possibility of unsuccessful evacuation due to inadequate training.
m.	Are there instructions about the means of escape?	Yes.	No.
n.	Are external escape routes properly maintained?	There are no external escape routes.	No.

6 Fire Warning and Detection

Item	Potential hazard	Details	Further assessment
a.	Can the existing detection system discover a fire quickly enough to raise the alarm in time for all occupants to escape to a place of safety?	Unsure. Also, not sure of arrangements made between employees of XYZ and that of the company who share the premises.	Yes. Assess for audibility of the alarm and assess joint evacuation problems.
b.	Can the means for giving warning be clearly heard throughout the premises?	See above.	Yes.
c.	Is there a back-up power supply for the fire warning system?	Yes.	No.
d.	Do employees know how to operate the fire warning system?	Yes. Simple break glass call points are provided.	No.
e.	Are there instructions for employees on how to operate the fire warning system?	Yes. All exit doors have signs fitted near them; these signs outline the action to take in the event of a fire.	No.
f.	Is the fire warning and detection system tested in accordance with appropriate British Standards and/or the manufacturer's instructions?	Yes. A fire safety company is contracted to service the alarm.	No.

7 Means of Fire-Fighting

Item	Potential hazard	Details	Further assessment
a.	Are fire extinguishers suitable for the purpose?	Unsure. The following extinguishers have been provided: Office area–one six-litre water extinguisher. Dispatch–one 2.5-kg dry powder extinguisher; one 3-kg carbon dioxide extinguisher; welding and grinding area–one 2.5-kg dry powder extinguisher.	Yes. Assess fire risks to determine whether there are enough items of fire equipment and if they are suitable.

(*Continued*)

Item	Potential hazard	Details	Further assessment
b.	Are fire extinguishers of suitable capacity?	See above.	Yes. See above.
c.	Are there enough extinguishers throughout the workplace?	See above.	Yes.
d.	Are extinguishers located in the correct positions throughout the workplace?	See above.	Yes.
e.	Can users gain access to fire extinguishers without exposing themselves to risk?	No. One of the extinguishers is located behind a storage cabinet at dispatch.	Yes. Assess risk to person who may use the fire extinguisher.
f.	Are the locations of fire extinguishers obvious?	Yes. All fire points have been fitted with signs that comply with the Safety Signs Regulations.	No.
g.	Have people been trained in the use of the fire extinguishers?	No.	Yes. Assess training needs of the employees.
h.	Are fire extinguishers tested in accordance with the manufacturer's instructions?	Yes. A contractor services them annually.	No.

8 Primary and Emergency Lighting

Item	Potential hazard	Details	Further assessment
a.	Is the primary lighting system adequate and functioning properly?	Yes. All areas of the workshop are adequately lit and the lighting is maintained on a regular basis.	No.
b.	Is the emergency lighting system tested in accordance with appropriate British Standards and/or the manufacturer's instructions?	Yes. A contract company tests the emergency lighting system annually.	No.

9 Fire Evacuation

Item	Potential hazard	Details	Further assessment
a.	Are there written procedures in place for evacuation in the event of a fire?	Signs are provided on all exit doors.	No.
b.	Have all personnel been trained in the procedures?	No.	Yes.
c.	Is there a system of training contractors in the evacuation procedures?	No.	Yes.
d.	Is there a visitor's register?	No. Numbers of visitors are usually very small–less than three at any one time.	No.

10 Fire Limitation

Item	Potential hazard	Details	Further assessment
a.	Is compartmentation provided within the premises?	No. None required by building regulations.	No.
b.	Is there any form of structural fire resistance within the premises?	No.	No.
c.	Is there a smoke ventilation system?	No.	No.
d.	Is there a pressurization system?	No.	No.

11 Fire Notices

Item	Potential hazard	Details	Further assessment
a.	Are notices provided on fire doors?	Yes. All fire doors are fitted with notices.	No.

12 Fire Safety Checks

Item	Potential hazard	Details	Further assessment
a.	Are checks carried out to ensure that windows and doors are closed when the premises are not occupied?	Yes, but informal.	Yes. Assess level of risk to determine security needs.
b.	Are checks carried out to ensure that electrical equipment is turned off and unplugged when the premises are not occupied?	As above.	Yes.
c.	Are checks carried out to ensure that smoker's materials are not left smouldering?	As above.	Yes.
d.	Are checks carried out to ensure that naked flames are extinguished when the premises are not occupied?	As above.	Yes.
e.	Are checks carried out to ensure that the premises are left secure when not occupied?	As above.	Yes.

13 Equipment and Plant

Item	Potential hazard	Details	Further assessment
a.	Are all items of electrical equipment, plant and machinery tested and maintained in accordance with the manufacturer's instructions?	Yes.	No.

14 Electrical and Gas Installations

Item	Potential hazard	Details	Further assessment
a.	Has the electrical system within the premises been tested in accordance with the Electricity at Work Regulations?	Yes.	No.
b.	Has the gas system within the premises been tested in accordance with the Gas Safety (Installations and Use) Regulations?	Not applicable.	No.

15 Liaison

Item	Potential hazard	Details	Further assessment
a.	Is liaison with the local fire authority maintained?	Annual visit by the fire brigade to check certificate.	No.
b.	Is there adequate access for fire appliances?	Yes. A large external yard joins the premises.	No.
c.	Are external assembly points adequately marked?	Yes. Signs are displayed in the car park.	No.
d.	Are fire hydrants and sprinkler valves adequately marked?	No hydrants or sprinklers installed.	No.

16 Emergency Plan

Item	Potential hazard	Details	Further assessment
a.	Is there an emergency plan?	No.	Yes. Assess need for an emergency plan.

17 Logbooks

Item	Potential hazard	Details	Further assessment
a.	Is there a logbook to record tests carried out on fire alarm and fire-detection equipment?	Yes.	No.
b.	Is there a logbook to record tests carried out on fire extinguishers and other fire-fighting equipment?	Yes.	No.
c.	Is there a logbook to record tests carried out on emergency lighting equipment?	Yes.	No.
d.	Is there a logbook to record fire training for employees?	No.	Yes.
e.	Is there a logbook to record tests on fire doors and associated equipment?	No.	Yes.

18 Legislation

Item	Potential hazard	Details	Further assessment
a.	What other legislation might be relevant to prevent fire within the premises?	Safety Signs and Signals Regulations. Highly Flammable Liquids Regulations.	Comply with the mentioned legislation by carrying out general health and safety risk assessments.

19 Review of Risk Assessment

Item	Potential hazard	Details	Further assessment
a.	Have arrangements been made to review the fire risk assessment at regular intervals?	No.	Yes. Assess when review is necessary.

APPENDIX B XYZ ENGINEERING LTD – RISK ASSESSMENT

Assessment number		
Assessor	Assessment date	Location assessed

Item number	Hazard	People at risk	Existing controls	Risk	Further controls
1.c.	Mobility-disabled persons becoming trapped by smoke.	One person works in the office area who is permanently confined to a wheelchair.	None.	4	Develop a procedure whereby another office employee takes care of the person in the wheelchair in the event of a fire. This 'helper' would guide and assist the person to a place of safety. Basic training should be provided for both persons.
1.f.	Arson.	Possibly employees, contractors and visitors; however, it is unlikely that arson would be committed during the day.	The premises are locked at night by the cleaning company employees. Windows are checked before the premises are locked.	6	At present, there is a considerable amount of wooden pallets, which are piled up against the wall at the rear of the premises. This should be cleared as soon as possible to avoid a risk of arsonists setting them alight.

(*Continued*)

Item number	Hazard	People at risk	Existing controls	Risk	Further controls
					A check should be made on the cleaning company to ascertain how many of their employees have access to keys. Also, the exact procedure for locking the premises should be checked and amended if necessary.
2.c.	Build-up of dust or other debris in the local ventilation system.	Employees.	The ventilation system was installed, to an appropriate BS, by XYZ employees.	6	It has been approximately 3 years since the LEV system was installed, during which time it has never been checked. Checks should be carried out by a competent person to ensure that airflow rates are sufficient to extract fumes and that the integrity of the unit is maintained. A contractor should be contacted and a contract established for the above work.

(*Continued*)

Item number	Hazard	People at risk	Existing controls	Risk	Further controls
2.f.	Combustible materials are stored above the dispatch area where vehicles operate.	Employees and contractors.	None.	6	All combustible materials should be removed from this area, as the risk of fire is significant. Only non-combustible materials should be stored in this area.
3.a.	Several employees smoke in both the office and the workshop area.	Employees, contractors and visitors.	Ashtrays have been provided in the office area; however, no such items have been provided in the workshop.	6	Owing to the high risk that smoking may present, especially in the workshop near the acetylene equipment, a no-smoking policy should be established.
3.b	Various items of portable electrical equipment are used in the premises.	Employees.	Although visual checks are carried out on portable electrical tools, these items of equipment are never formally inspected or checked.	4	All items of portable electrical equipment should be checked on a regular basis to ensure that they continue to operate safely. There are two options to achieve this. Firstly, a contract could be established with an electrical company to check all items of equipment annually.

(Continued)

Item number	Hazard	People at risk	Existing controls	Risk	Further controls
					Secondly, a portable appliance tester could be purchased and the checks carried out in-house. Owing to the risk, which is moderate, it is perhaps better to take a less costly option to control the risk of fire from these items. Therefore, a portable appliance tester should be purchased and all items of electrical equipment should be tested annually.
3.e.	Naked flames.	Employees and contractors.	None.	9	Combustible material was found near the area where acetylene equipment is normally used. Naked flames are used to ignite the acetylene torch. All combustible materials should be removed from this area.

(*Continued*)

Item number	Hazard	People at risk	Existing controls	Risk	Further controls
3.f.	Vehicles.	Employees and contractors.	A fork-lift truck operates within the dispatch area, which is driven by a fully trained and competent person. However, no checks are carried out on the truck.	3	Introduce a daily pre-start check for the truck. A checklist should be developed to allow the driver to systematically check all important areas of the truck.
3.l.	Machinery.	Employees.	Bench grinders are used within the workshop.	3	Regular tests and maintenance should be carried out on the bench grinders.
3.o.	Hot work.	Employees.	The acetylene equipment is regularly maintained by a contract company.	6	An additional carbon dioxide fire extinguisher should be provided near the area where the acetylene equipment is used.
3.p.	Arson.	Employees, contractors and visitors.	None.	9	All combustible materials should be removed from external areas of the site. Skips should be removed regularly to ensure that they do not overflow with waste. Additional external lighting should be installed to ensure that all areas of the site are adequately illuminated.

(*Continued*)

Item number	Hazard	People at risk	Existing controls	Risk	Further controls
3.q.	Arson.	Employees, contractors and visitors.	None.	9	As above.
4.b.	Combustible material on escape routes.	Employees and contractors.	None.	4	Daily checks should be carried out to ensure that combustible waste does not accumulate on escape routes.
4.f.	Combustible waste.	Employees and contractors.	None.	4	As above.
5.d.	Means of escape.	Employees, contractors and visitors.	None.	9	Fire exit doors are never checked for their operation. The door in the middle of the workshop should be repaired to ensure it opens freely. Regular inspections should be carried out to ensure that all fire doors are operating properly.
5.f.	Means of escape.	Employees, contractors and visitors.	None.	4	As above.
5.l.	Training of employees.	Employees.	None.	6	No training has been provided for employees about the action to be taken in the event of a fire. Suitable fire drills should be arranged and carried out every six months.

(*Continued*)

Item number	Hazard	People at risk	Existing controls	Risk	Further controls
6.a.	Fire alarm.	Employees, contractors and visitors.	None.	6	The fire alarm should be tested at regular intervals. Because the premises are being shared by another employer, regular joint fire evacuations should take place between employees from both companies.
6.b.	Fire alarm.	Employees, contractors and visitors.	None.	6	An audibility test should be carried out to ensure that the fire alarm can be heard in all areas of the premises.
7.a–d.	Fire extinguishers.	Employees.	Several fire extinguishers are provided at present and are checked annually by a competent contractor.	6	An additional fire extinguisher should be provided at the dispatch area; for vehicle engine fires – one 5-kg carbon dioxide fire extinguisher.
7.e.	Fire extinguishers.	Employees.	None.	4	Relocate the extinguisher that is presently located behind a storage cabinet at dispatch.

(*Continued*)

Item number	Hazard	People at risk	Existing controls	Risk	Further controls
7.g.	Fire extinguisher training.	Employees.	None.	6	Low-cost training should be provided for workshop employees on the proper technique of using fire extinguishers. This training should be provided as a refresher course in three years.
9.b,c.	Fire evacuation.	Employees and contractors.	No training provided.	4	Provide regular evacuation training for employees and contractors.
12.a–e	Fire safety checks.	Employees.	Informal checks are carried out at the end of each working day.	6	A formal procedure should be developed to ensure that all windows are closed before people leave the premises. Also, the procedure should cover the following: electrical equipment turned off; smoker's materials are extinguished; naked flames are turned off; and all external doors are locked.

(*Continued*)

Item number	Hazard	People at risk	Existing controls	Risk	Further controls
16.a.	Emergency plan.	Employees and others.	None.	4	An emergency plan should be compiled jointly with the employer who shares the premises. The emergency plan should detail maintenance of the means of escape; raising the alarm; and contacting the emergency services.
17.d,e.	Logbooks	Employees.	Some logbooks are currently kept for testing of fire equipment.	3	Logbooks should be kept for the following tests: fire alarm and fire- detection equipment; fire extinguishers; emergency lighting tests; fire training and fire drills; and repairs on fire exits.
19.a	Review of fire assessment.	Employees.	None.	6	Owing to the number of action points as a result of this fire risk assessment, the assessment should be reviewed in six months.

(*Continued*)

Other applicable regulations	Details
Safety Signs and Signals Regulations	Carry out a survey of all signs to ensure that they comply with the regulations.
Highly Flammable Liquids Regulations	Check all storage containers to ensure that they comply with the regulations.

APPENDIX C XYZ ENGINEERING LTD – ACTION PLAN

Assessment number		
Assessor	Assessment date	Location assessed

Item number	Action	Responsibility	Target date	Completion date
1.c.	Develop a procedure for dealing with wheelchair-confined employees in the event of a fire.		6 months.	
1.f.	Clear pallets and debris from the yard.		2 weeks.	
1.f.	Carry out a check to ascertain who has the keys to the premises, including employees and contractors. If necessary, change locks and limit the number of keys in circulation.		2 weeks.	
2.c.	Contact a contractor to check the local exhaust ventilation (LEV) system.		2 weeks.	
2.c.	Develop a procedure whereby employees carry out a daily check of the main components of the LEV system.		1 month.	
2.f.	Remove combustible material from the storage area above dispatch.		2 weeks.	
3.a.	Establish a no-smoking policy for all employees, contractors and visitors.		1 month.	
3.b.	Purchase a low-cost portable appliance tester.		2 months.	
3.b.	Develop a procedure to have all portable electrical equipment tested annually.		2 months.	
3.e.	Remove combustible material from the area where acetylene is used.		Immediate.	

(*Continued*)

Item number	Action	Responsibility	Target date	Completion date
3.f.	Develop a procedure for checking fork-lift truck on a daily basis.		4 months.	
3.l.	Introduce a maintenance programme for the bench grinders.		4 months.	
3.o.	Provide a carbon dioxide fire extinguisher.		2 weeks.	
3.p. and 3.q.	Remove waste material from external areas.		Immediate.	
4.b. and 4.f.	Ensure daily checks are carried out to ensure escape routes are not blocked.		2 weeks.	
5.d. and 5.f.	Repair the fire exit door from the workshop.		1 week.	
5.l.	Develop a procedure whereby all employees and contractors regularly undertake fire drills.		1 month.	
6.a.	Ensure that the fire alarm is tested on a regular basis.		1 month.	
6.b.	Carry out an audibility test of the fire alarm to ensure that it can be heard throughout the premises.		1 month.	
7.a.	Provide an additional fire extinguisher at the dispatch area.		1 month.	
7.e.	Relocate the fire extinguisher.		2 months.	
7.g.	Provide training for employees in the use of fire extinguishers.		2 months.	
9.b,c.	Provide regular evacuation training for employees and contractors.		6 months.	
12 a.–e.	Develop a procedure whereby checks are carried out on the premises before people leave at the end of each day.		2 months.	
16.a.	Develop a joint emergency plan.		6 months.	
17.d,e.	Develop a procedure whereby logbooks are completed for all tests of emergency equipment and training.		1 year.	
19.a.	Review the fire risk assessment.		6 months.	

4 Fire Safety Management

'FIRE CAUSES STRUCTURAL DAMAGE TO FACTORY DURING REFURBISHMENT'

A metal refining company was fined £15 500 after a fire caused structural damage to a part of its factory.

The refurbishment work involved the use of welding equipment to remove gas pipes from the walls of a storage area. However, bags of sodium chlorate, which is a flammable substance, were left in the storage area while work was carried out.

During the work, which was carried out by a contractor, sparks from the cutting operation landed on the bags of chemicals and ignited the substance. The fire spread quickly through part of the factory.

The court found that the company had failed to carry out an adequate fire risk assessment, which would have identified the need to remove the bags of chemicals before cutting operations began. In addition, the contractors working for the company should have been informed of the risks of cutting near the bags.

'MAN KILLED AFTER A CAN OF SOLVENT EXPLODES'

An engineering company was fined £14 000 after one of its employees was killed. The man had been engulfed in flames after using a can of solvent to start a fire.

Wooden pallets were placed inside a metal cage and then ignited. Common practice involved the use of solvents to speed up the ignition process. Operators saturated the pallets with the solvents before applying a flame to ignite the material.

It was proposed, on this occasion, that the employee had ignited the pallets and then placed the can of solvent on top of a nearby cardboard box. It is likely that the fire spread to the box, causing the can to explode, which ignited his clothes. The employee suffered extensive burns and later died in hospital.

Although the company had carried out risk assessments for engineering activities, it had failed to carry out fire risk assessments for the task of burning pallets. As a result of this failure, no safe system of work had been developed for the task.

'MANAGING DIRECTOR FINED AFTER EMPLOYEE IS KILLED'

A Managing Director of a petroleum decommissioning company was fined £10 000 when a court found that an employee was killed after an underground tank exploded. The company failed to develop safe working practices for work involved with decommissioning petrol stations.

Welding equipment was used to dismantle metal supports above a disused petrol tank. A bead of hot metal from the supports fell down into an uncovered manhole that led to the underground tank. The bead ignited petrol vapour in the tank, causing a massive explosion.

The company had failed to have proper working practices in place and was fined for breaching Section 2 and 3 of the Health and Safety at Work Act 1974.

'COMPANY FINED £300 000 AFTER WORKER DIES IN AN EXPLOSION'

In sentencing the company, the judge told the court that a 29-year-old employee was killed because the company had in place an 'old-fashioned', complacent and wholly inadequate approach to risk assessment.

The employee had been working in an area where cylinders were filled with toxic and flammable gases. The gas was stored in bulk tanks and transferred into individual cylinders in accordance with the manufacturer's instructions. However, during a particular filling operation, the employee had mixed the gases incorrectly, creating a highly sensitive pressurized gas that, on opening the cylinder valve, caused the substance to ignite.

Inevitably, the gas cylinder exploded, killing the employee and causing extensive fire damage to the site.

The company had failed to carry out an adequate risk assessment, had failed to take into consideration the employee's lack of experience and had failed to provide proper training in the safe working practices that are necessary for the task.

'LAD BURNED ON FIRST DAY AT WORK'

On his first day at work, a 17-year old lit a cigarette while in his firm's delivery van, which contained drums of white spirit. Seconds after, vapours ignited, engulfing him in flames and causing severe burns to his legs.

After completing a task, the young employee climbed into the van and immediately lit a cigarette. One of the cans of white spirit had been opened and some of the chemical had spilled onto the employee's clothing. Lighting the cigarette had caused the vapours around his clothing and that around the open container to ignite.

The company had failed to assess fire risk properly, had failed to provide fire extinguishers inside the van, even though it was used to transport flammable material, it had failed to provide 'no-smoking' signs, and had failed to provide adequate training for employees.

The company was fined £15 000 and was ordered to pay costs of more than £1000.

The above examples illustrate the importance of proper fire risk assessments and proper control measures. This chapter looks at the various types of management controls that may be necessary to control fire risks in the workplace.

Generally, control measures can be divided into three categories:

- Hardware controls – This type of control involves items such as fire doors, fire extinguishers, sprinkler systems, hose reels and other fixed systems and fire safety signs. Hardware controls are often required for dealing with fire after it has started or for limiting fire, heat and smoke spread. Hardware controls are often referred to as reactive control measures – reacting after the event to allow occupants to escape safely.
- Management controls – Often, hardware alone is not enough to control the risk from fire. Management controls include procedures, job instructions, safe systems of work – all of which are methods for ensuring that employees carry out work in a prescribed and safe manner. Also under the category of management controls are maintenance procedures for equipment, plant and machinery. One important method of preventing fire from starting is to ensure that equipment is properly maintained. Finally, under this category is regular inspection of the workplace and working methods.
- Training – Training of employees may be provided in the form of simple fire drills held at regular intervals. For more complex operations, training may have to be much more defined, with specialized roles to perform in emergency situations.

Some of the cases mentioned previously highlight the need to ensure that employees know exactly what is required of them. Management must ensure that employees have the capability and level of knowledge to carry out such tasks.

Depending on the complexity of the activities carried out, employees may also require basic training in the use of fire-fighting equipment and training in evacuation techniques.

HARDWARE CONTROLS

Fire-Fighting Equipment

It is important that the correct type and number of fire extinguishers are strategically placed throughout the workplace. If the premises have an existing fire certificate, it is likely that there will be enough extinguishers. However, depending on the results of the fire risk assessment, additional fire extinguishers may be required, especially for specific risks.

As an aid to help identify which extinguishers are required for a given hazard, it is useful to consider British Standard EN2, which classifies fire as follows:

- Class A Fires involving solid materials.
- Class B Fires involving liquids or liquefiable solids.
- Class C Fires involving gases.
- Class D Fires involving metals.

The best method of protecting a Class A fire risk is to provide a water-type extinguisher. Foam and dry powder fire extinguishers are also useful.

For a Class B fire risk, the best type of extinguisher is foam, although carbon dioxide or dry powder types can also be used.

Dry powder extinguishers can cover Class C fire risks. However, the first line of defense if a fire occurs in a gas is not to extinguish it but to isolate the supply. It is important to remember that if a gas fire is extinguished, it may result in escaping gas, which is often colourless and heavier than air. This produces a greater risk of subsequent explosion.

Class D fire risks are generally rare, involving materials such as aluminium, magnesium or sodium. None of the above fire extinguishers can be used successfully to extinguish a Class D fire. Special powders are required to deal with these fires and are available from specialized stockists.

After carrying out a detailed fire risk assessment, it may be necessary to provide some, or all, of the above types of extinguisher. The assessment should identify how many extinguishers are required for a specific risk and what types are required. For example, assume that a room is used for storing mechanical equipment and that in the corner of the room there is a large electrical switch panel. The fire risk assessment might identify that, should a fire start at the panel, a single carbon dioxide-type extinguisher would be suitable to deal with the fire. Therefore, as a result of the assessment, one carbon dioxide extinguisher should be purchased and placed at a safe distance from the panel.

It must be remembered that each type of fire extinguisher requires a different technique to extinguish a fire successfully. The fire risk assessment should identify who is most likely to be present if a fire is detected. Suitable training should be provided in the particular technique that is used for fighting the fire.

Fire extinguishers are not only needed to fight small fires in the workplace but are also used as a means of protecting the 'means of escape'. If people are evacuating from a building and they find that the fire has breached their escape route, it might be necessary to use an extinguisher for a short time to assist evacuation. Once again, this should be identified in the risk assessment, and suitable training in the general use of fire extinguishers should be provided for employees.

Fire-Detection Systems

To ensure that people are able to escape from the premises, it is essential that some form of fire detection is provided, especially in areas that are normally not occupied.

In small premises, it might be sufficient to provide inexpensive domestic-type smoke detectors. In small premises, it would not be a justifiable expense to install complex fire-detection equipment. However, it is important to provide cover for certain high-risk areas, even in small premises.

These high-risk areas include rooms that are normally not occupied, for example, store rooms and basements. In order to evacuate the premises, it is essential that these areas be covered with fire detection, even in the smallest of premises.

Another consideration is the number of escape routes from an area of the premises. If a single route of escape is available from a room, via another room, it is essential that a smoke detector be installed to provide an early warning of fire.

Simple smoke detectors are a very effective way of raising the alarm; however, they are only as effective as long as they work. Care should be taken to ensure that batteries are replaced on a regular basis and weekly tests are carried out to test their efficiency.

In larger, more complex premises, especially where sleeping accommodation is provided, more complex forms of fire detection will be necessary. Depending on the results of the assessment, fire-detection equipment may need to include control panels, automatic links to fire-fighting equipment and perhaps even automatic links to collector stations, which in turn notify the local fire brigade.

Fire Warning System

Once a fire has been detected, it is essential that all people within the premises be notified that a fire has started. In its simplest form, one employee can shout, 'FIRE', which might be sufficient to notify everyone else within the premises. However, this relies on the fact that the premises are small and that everyone works within the same area.

When employees, visitors and contractors, occupy a building and are spread throughout the building, shouting 'fire' is not going to be an effective method of notifying everyone. In this case, a more complex system may have to be installed.

An electrically operated fire alarm is a reasonably inexpensive method of raising the alarm throughout the premises and usually consists of a series of manual call points and sounders placed strategically throughout the building.

Call points may be required along escape routes (depending on the length of the corridors) and may be required adjacent to final exit doors and at the head of the staircases on each floor. Careful placement of call points will ensure that employees escaping from a building will hopefully raise the alarm as they escape along a corridor or when they reach a final exit.

The fire alarm should emit a distinctive sound, quite different from any other sound used to notify people, e.g. clocking in or out time alarms. It is essential that all employees can hear the fire alarm, especially when they are in areas that are usually unoccupied. It is important that fire alarms are tested and all places within the building are visited to ensure that the alarm could be heard.

Another important aspect is to ensure that fire alarms can be heard above the background noise. For example, in a busy workshop where there may be several items of machinery in use, the background noise could well be in excess of the level emitted from an alarm sounder. In this case, the fire risk assessment would identify this as a potential problem. From there, a suitable solution should be identified; in this case, perhaps an additional sounder would be required near particularly noisy equipment.

Another consideration is that employees, visitors and contractors may be wearing ear protection because of high factory noise levels. It is essential that sounders can be heard through the protective equipment. In extreme cases, it may be necessary to install visual warning devices such as flashing lights, which are linked to the alarm system.

A fire risk assessment should also consider the type of employees that are present. Some companies employ people with hearing disabilities. In this case, it might be necessary to install visual warning devices or, as an alternative, develop a procedure whereby other employees guide disabled persons to safety.

In larger buildings, it might be necessary to install more complex systems, particularly where members of the public are involved, e.g. shopping centres and theatres. In such places, public address systems may also be necessary to help direct people in the event of a fire.

Escape Routes

Part of the fire risk assessment process is to ensure that employees, visitors and contractors can escape to a place of safety within a reasonable length of time. This might be quite simple in small, single-storey premises, and in this case, no structural protection may be necessary for escape routes. However, in much larger premises, perhaps one where there are many floors, escaping from the top floor via stairways and corridors might take a considerable time. Some form of structural protection would be required to protect the integrity of the escape route to allow persons to escape to a place of safety. This protected route is classed as a 'place of relative safety', which then leads to a final exit. A place of relative safety is one which is protected by structural fire-resisting material and will maintain its integrity for enough time to enable people to escape to safety.

One of the most important aspects of a fire risk assessment is to consider how far a person must travel to get to a place of safety, which is usually defined as 'a place beyond a building where a person is no longer in danger from fire'.

Assessing travel distances to a place of relative safety or to a place of safety can be a complex task. If in doubt, and especially before any major structural alterations are performed on a building, advice should be sought from the local fire authority or from the local building control department.

However, the following guidance is helpful when carrying out an initial fire risk assessment to determine whether there may be a problem with travel distances.

To carry out this assessment, it is first necessary to categorize the premises as low, medium or high risk.

Low-Risk Premises

Low-risk premises are usually those where there is minimal risk to people's lives and where the risk of fire is low.

Medium-Risk Premises

Medium-risk premises can generally be classed as those where any outbreak of fire is likely to remain confined to a particular area, allowing people to escape quickly. Generally, medium-risk premises are those where the number of occupants is small, with few people requiring specialized assistance to escape from the premises.

High-Risk Premises

High-risk premises are those where large numbers of people occupy the premises or where sleeping accommodation is provided.

High-risk premises might be those where, owing to the nature of activities, there is a high potential for a fire to start. In addition, there may be highly flammable materials stored or used within the premises.

Table 4.1 can be used to assess travel distances. The table is divided into two sections. The first section applies to premises where there is only one route of escape from the premises. The second section applies to premises where there is more than one route of escape.

Table 4.1

Available escape routes	Type of fire risk	Maximum distance to travel to a place of safety
One	High	12 m
	Medium (sleeping accommodation provided)	16 m
	Medium	18 m
	Medium (factories)	25 m
	Low	45 m
More than one	High	25 m
	Medium (sleeping accommodation provided)	32 m
	Medium	45 m
	Low	60 m

Fire Escape Doors

Fire escape doors need to be wide enough to allow people to pass through in the event of an evacuation. Several factors must be taken into consideration when assessing the adequacy of fire doors. First, fire escape doors should open in the direction of travel where:

- more than 50 people may have to use the door;
- doors are at the foot of a stairway;
- doors are situated within a high-risk area;
- doors are on an escape route that is occupied by members of the public.

Second, all final exit doors leading from the building to external areas should be easily opened without the use of a key. In case these doors have to be secured to prevent people from opening them from outside, the doors should be fitted with a single form of release device such as a turn button, panic bolt, or push pad.

Emergency Exit Signs

All fire safety signs must comply with the Health and Safety (Safety Signs and Signals) Regulations 1996. Signs are used to indicate escape routes and emergency exits.

Lighting

Escape routes must be lit properly to allow occupants to escape safely at any time of day or night. Arrangements should be made to simulate a primary lighting failure to assess the efficiency of emergency lighting. Additional lighting may be required or maintenance work may be necessary on the existing system to improve its illumination qualities.

Emergency lighting should:

- clearly indicate all escape routes,
- clearly indicate fire-fighting equipment, such as extinguishers, fire blankets and so forth,
- clearly indicate fire alarm call points.

While carrying out a fire risk assessment, it is important to determine whether 'danger' areas are adequately lit by emergency lighting. These areas include the following:

- points where corridors meet
- exit doors
- any change in floor level
- tops of staircases
- any hazardous equipment, machinery or processes
- externally at all final exits.

MANAGEMENT CONTROLS

Fire Safety Policy

Modern health and safety management relies on the development and implementation of effective policies. Health and safety policies are an essential part of the risk management process and should consist of three main elements:

- A general statement of intent.
- A section detailing roles and responsibilities.
- An arrangements section, which should contain more detailed information relating to procedures that should be adopted to maintain a safe place of work.

Like general health and safety management, it is good practice to develop a fire policy. To be effective, a fire policy should be communicated to all employees and contractors who frequently use the premises. In addition, other people such as visitors might also need to be aware of the policy.

A fire policy should consist of the following elements:

General Statement of Intent

This should be signed by the most senior person in the company and should clearly state that all employees regardless of their status should work together to ensure that the workplace remains 'fire-safe'.

Developing such an environment will not happen overnight. It takes time to develop a culture where the prevention of fire is considered by everyone in the organization to be of high priority.

A statement of intent can also outline goals, aims and objectives of the company as far as fire safety management is concerned. As a result of the fire risk assessment process, many additional control measures may have been identified. It would be impractical for a company to put each control measure in place immediately. Instead, an action plan may have to be developed to programme in control measures, perhaps in conjunction with financial planning. These action points can be turned into aims and objectives, which can then be published as part of the statement of intent.

The above approach to action planning is a very effective method of showing employees that management is committed to ensuring that the workplace is kept as safe as possible. However, it does rely on action being taken. There is little point in publishing objectives and action plans if nothing is actually done and no action is carried out.

The general statement of intent should be updated regularly and displayed throughout the premises.

Responsibilities

The second part of a fire policy should contain information relating to responsibilities for fire safety. It is everyone's responsibility to ensure that the workplace is kept safe (Health and Safety at Work Act – Employees Duties); however, there are certain additional responsibilities that may need to be detailed within the responsibilities section.

In a large organization, it might be necessary to appoint a person to look after fire matters. This person might be a properly trained and experienced fire professional. However, most organizations will not have to appoint dedicated fire officers; instead, an existing employee, probably a line manager or senior manager, can be given responsibility for fire safety, and his or her roles and responsibilities should be outlined within the fire safety policy.

It is often necessary to appoint fire marshals or specific employees to assist disabled persons in the event of a fire. Once again, these people should have their responsibilities outlined in the policy. They should be provided with suitable training to ensure that they can carry out their duties in an efficient and safe manner.

Arrangements Section

Finally, the fire safety policy should contain a section relating to specific fire procedures. This section might include procedures such as:

- safe shutdown of plant in the event of fire,
- evacuation of premises,
- evacuation of specific employees, including disabled persons,
- controlling contractors in the event of fire,
- maintaining security after fire evacuation,
- notifying the fire brigade that a fire may have started,
- first-aid fire-fighting arrangements,
- training for fire marshals, general evacuation training and more specific training, if necessary.

Maintenance and Testing of Equipment

A major cause of fires in the workplace is due to poor maintenance of plant and equipment. Often, it is necessary to put in place a planned maintenance programme to ensure that the plant and equipment is maintained in accordance with the manufacturers' instructions.

Fires most commonly start in plant and equipment because of the following contributory factors:

- Overheating of electrical equipment due to dust build-up of ventilation points.
- Overheating of electrical equipment due to rubbish or fabric items blocking ventilation points.
- Loose drive belts.
- Bearings that are not lubricated properly.
- Faults in electrical equipment or breaks in electric cables.
- Continual misuse and bad handling of portable electrical equipment.

- Flammable materials too close to exposed hot surfaces where insulation has broken down.
- Leaks from valves, pipes or flanges.
- Poor earthing of electrical equipment.
- Static electricity.

The Provision and Use of Work Equipment Regulations require employers to ensure that equipment is properly maintained and subjected to regular inspection. This provides an ideal system to ensure that all equipment with the potential to start a fire is checked on a regular basis and proper maintenance procedures followed.

Testing of Emergency Equipment

All fire-fighting, fire-detection and fire warning equipment must be tested regularly to ensure that they are in efficient working order should a fire occur. A regular check of all escape routes, fire doors, emergency lighting and fire safety warning signs needs to be carried out. Any defects found should receive attention as quickly as possible.

It is often a good idea to carry out weekly 'walk-through' inspections of the entire workplace. This can be carried out by anyone in the organization, and an effective method is to alternate the type of person carrying out the inspection. For example, one week the inspection could be carried out by a senior manager, while the following week an administrative assistant could do the inspection. Provided the checklist is well detailed, the task does not have to be undertaken by those with specific knowledge of fire safety. This approach helps to promote 'ownership' of health and safety among all employees.

The contents of a walk-through type inspection might include the following:

- Are escape routes blocked or restricted by rubbish or other debris?
- Are fire extinguishers provided in the correct positions?
- Have fire extinguishers been removed from their holders?
- Are fire safety signs clean and clearly visible?
- Are fire doors wedged open or difficult to open?
- Are final exits difficult to open?
- Are any flammable materials stored in escape routes?
- Are fire notices clearly displayed in their correct positions?
- Are all self-closing devices operating properly?
- Are all panic-type fastenings operating properly?

In addition to visual inspections, fire equipment should be tested regularly in accordance with the appropriate British Standards. As a general guide, Table 4.2 illustrates the minimum frequency of tests that are necessary to ensure that the equipment functions properly.

Table 4.2

Type of equipment	Check or test	Frequency
Fire-detection equipment	Test operation of self-contained detectors and manually operated devices.	Weekly
Fire warning equipment	Test operation of alarm and the audibility of sounders (ensure sounders can be heard from all areas of the workplace). Each week, a different break glass call-point should be used to set the alarm off.	Weekly
Fire-fighting equipment	Check all fire extinguishers to ensure that they are located at the proper points. Visual checks to ensure that they have not been tampered with.	Weekly
Emergency lighting equipment	Check all systems to ensure that they operate properly.	Monthly
Fire-detection equipment	Full check and test of the system by a competent service engineer.	Annually
Fire warning equipment	A full check and test of the system by a competent engineer.	Annually
Emergency lighting equipment	A full check and test of the system by a competent engineer.	Annually
Fire-fighting equipment	A full check and test of each fire extinguisher by a competent person.	Annually

Fire Evacuation Procedures

Depending on the results of the fire risk assessment, it will be necessary to evacuate the premises in the event of a fire in most cases. A fire evacuation procedure should be developed and must take into consideration the following:

- Will the premises be completely evacuated in the event of a fire alarm sounding or will there be a phased evacuation?
- Are there employees with special needs to be considered during fire evacuation?
- How will a role call be taken?
- Is it necessary to provide fire marshals instead of taking a roll call?
- Are emergency shutdown procedures necessary?
- Are refuge areas necessary for wheelchair-dependent employees?

Complete Evacuation or Phased Evacuation

In small to medium-sized premises, it is probably better to evacuate the whole premises if a fire alarm sounds. However, in larger premises, it might be more appropriate to evacuate in phases. This will depend on the number of occupants and the type of people who normally occupy the premises, e.g. members of the public. Phased evacuation is a complex subject and may be dependent on other existing fire precautions within the premises. For this reason, the local fire authority should be contacted if phased evacuation is contemplated.

Disabled Employees

The Disability Discrimination Act 1995 requires employers to make 'reasonable adjustments' to their premises to ensure that no employees are at a disadvantage. This includes ensuring that disabled employees and visitors can escape from the premises in a safe manner.

The fire risk assessment will help to identify the likelihood that disabled persons may be present within the workplace.

One of the major problems associated with the evacuation of disabled people is that some may be confined to wheelchairs. Obviously, this presents a problem where staircases are used as part of the fire escape route. In order to protect wheelchair-dependent persons, it might be necessary to provide refuge areas. These are specially designated areas on all floors that are constructed from fire-resisting material. Fire refuge areas are designed to maintain their integrity for a specified period of time. Wheelchair-bound people, and other disabled persons, move to these areas, where they will be safe from fire and await further assistance to evacuate.

In some cases, it might not be reasonably practicable to construct refuge areas. In small premises where there is more than one floor, an alternative is to arrange for manual evacuation. In this case, other employees with specific responsibility would be required to assist disabled persons to evacuate.

Once again, the fire risk assessment will help to establish the nature of risk and will help to ascertain what is reasonably practicable in terms of solving the problem.

Another group of employees with special needs are those with impaired vision. Some people have difficulty in seeing colour safety signs and some have problems in reading print. This should be taken into consideration when assessing the risk from fire and may require additional larger signs to be provided. Further advice on this may need to be sought from appropriate disability organizations.

General Housekeeping

A significant cause of workplace fires is the build-up of flammable material, including general rubbish, packaging materials, empty flammable material containers and waste that accumulates as a result of construction work. It is essential that such waste materials be regularly removed from the premises.

Regular, daily checks should be carried out on premises to ensure that rubbish and other flammable materials are not stored in escape routes or near fire-fighting equipment.

Particular areas that should always be checked regularly include storerooms, basements and attics. This is particularly important due to the fact that these areas are usually unoccupied for long periods of time. Any rubbish found in these areas should be removed promptly.

It is not only internal areas that need to be kept clear of rubbish. Often, when arson attacks are made to the premises, the perpetrators will use discarded flammable waste to ignite a fire, which then spreads to the building itself. It is important that inspections made internally are extended to external parts of the premises. Any accumulation of rubbish against external walls should be removed promptly. It is good practice to mark on the ground areas where rubbish should not be stored.

Skips are another source of concern. Skips positioned too near a building may spread a fire quickly. Thousands of skip fires occur every year, some started deliberately and others by employees accidentally discarding hot materials.

Many serious fires occur during building and maintenance work. This type of activity needs to be monitored very closely and will certainly require further fire risk assessments.

Particular attention should be made to the following:

- Accumulation of rubbish and other flammable waste.
- Equipment, tools, building materials and general waste blocking fire escape routes, fire doors and fire exits.
- Use of fire extinguishers to wedge open fire doors.
- Use of other devices to disable fire doors.
- Breaches in fire-resisting walls, floors, partitions and other parts of the building structure.
- Portable tools that introduce a risk of electrical-type fires.
- Disarming fire alarms and detection systems.

During construction work, it is important to conduct regular checks of the premises on a daily basis. All contractors employed should be made aware of the fire arrangements and the needs of any particular employees. This will help ensure that successful evacuation can occur regardless of the type and scale of the construction or maintenance work.

Storage Arrangements for Highly Flammable Gases and Liquids

Highly flammable materials will be dealt with later; however, some general precautions are needed when storing or using such substances. The following general precautions should be followed:

- Quantities of flammable and highly flammable materials should be kept to a minimum in the workplace.
- Flammable materials and highly flammable materials should be kept clear of escape routes, fire doors, fire exits and fire-fighting equipment.
- As far as possible, highly flammable materials should be replaced with other less flammable substitutes.

- Appropriate fire-fighting equipment should be located near places where flammables are stored and used.
- All employees using flammable materials should be given appropriate training in their use and how to deal with emergencies.
- Quantities of flammables not in use within the workplace should be stored externally in secured cabinets or internally in fire-resisting stores.
- Hot surfaces or other ignition sources should be kept away from storage areas used for flammable materials.
- Contractors should be made aware of the precautions in place for storing and using flammable materials.
- Decanting of flammable materials should be undertaken in the open air, using appropriate containers.
- Personal protective equipment should be issued to employees.

Safe Systems of Work

As a result of the fire risk assessment, it may be necessary to develop procedures for certain tasks. If a task or activity has been identified as presenting a significant risk of fire, a documented safe system of work will be required. Safe systems of work are like job procedures and are used by employees to complete a task in a standard manner. It is therefore essential that employees carrying out these tasks be given proper instruction in the methods to adopt.

In certain cases, a more formal safe system of work may be required. For example, hot work such as welding, flame-cutting, the use of blowlamps or grinding equipment – all present a significant risk of fire when used near combustible materials. Even the simple task of stripping paint from a window surround with a blowlamp could start a fire, with possible serious consequences.

For such work, a standard procedure may not be enough to control the fire risk. In these cases, a permit to work would probably be better to ensure that all fire hazards are removed or controlled.

Permit-to-work systems ensure that:

- there is a formal check confirming that a procedure is being followed;
- coordination is taking place between employees and contractors;
- time limits are set for the work to be carried out, and where these time limits cannot be achieved, further permits need to be issued;
- identify specialist equipment for fire-fighting if required;
- identify specialist personal protective equipment for employees;
- identify other arrangements that may have to be put in place.

A more detailed description of the permit-to-work system is included in later chapters.

Emergency Planning

Should a fire occur within the workplace, employees, contractors and visitors need to know what to do. Also, should special arrangements need to be made to shut down plant or to help other people escape, people need to know what the arrangements are.

An emergency plan should be developed and tested regularly to ensure that everyone knows what to do in the event of a fire or other emergency.

An emergency plan will also ensure that the premises is left as safe as possible and will help minimize any damage as a result of fire.

For simple workplaces, the emergency plan may be no more than a simple written instruction posted strategically throughout the premises. Employees would be able to see these notices and respond accordingly, provided they had received sufficient training.

However, in more complex situations, the emergency plan may need to take into account the following:

- General evacuation by all people or phased evacuation.
- Essential personnel such as technicians and engineers to remain to shut down plant and equipment.
- Arrangements for a first-aid fire-fighting attack.
- Arrangements for a roll call to be taken after evacuation.
- Fire marshals to check that the premises have been evacuated.
- Assistance for employees with special needs.
- The method of calling and notifying the fire brigade.
- Arrangements for informing the fire brigade that hazardous materials are stored or that hazardous equipment may be present within the premises.
- Minimizing loss should a fire occur.
- Notifications, should a serious fire occur, e.g. insurance company, customers etc.

Depending on the nature of risk, it may be necessary to provide drawings of the premises. This will help employees and may also be of assistance to the fire brigade. Such drawings might include the following:

- Locations of fire-fighting equipment.
- Locations of stop valves for fixed fire-fighting equipment, such as sprinklers, deluge systems or bulk discharge units.
- Locations of fire escape routes.
- Locations of fire refuge areas.
- Assembly points.
- Location of the main electrical supply and gas isolation points.
- Location of any storage areas for flammable materials.
- Locations of any hazardous items of equipment or plant.
- Locations of any hazardous structural features such as unguarded pits, hot surfaces, electrical switchgear and so forth.

The emergency plan should be tested regularly whereby all employees take part in regular fire drills and full evacuations. If the premises are particularly complex, it might be necessary to involve the local fire brigade in fire drills to test how their response integrates with the company's plans.

Training

Information and Instruction for Employees

It is important that all employees, including those who do not normally occupy the premises during normal working hours, are provided with adequate information, instruction and training in the risks from fire and the action to take in the event of a fire occurring.

Information usually takes the form of fire notices, fire safety policies, procedures to be followed while carrying out certain activities and permit-to-work systems. It is important that the information is provided to employees in a simple manner. The more complex the instructions are, fewer will be the number of people who would be able to follow them. Fire notices should be written in simple and direct language and should clearly outline the procedure to follow in the event of a fire. It is important to remember any visually impaired employees or people who do not speak English.

Instruction may take the form of an employee being shown around the premises and being told of any hazardous areas and the action to take in the event of a fire. Instruction includes following the escape route to a final exit and pointing out the locations of fire-fighting equipment. On the first day of employment, new employees should be given instruction on the following:

- The location of escape routes, fire doors and the method of operating final exit doors.
- The method of raising the alarm.
- Assembly areas.
- Not to use lifts unless they are specially constructed for use in fire conditions.
- Not to return to the premises until given permission.

Training should be provided in the form of regular fire drills and fire evacuation. During such occasions, employees, and others, should be made aware of the emergency plan.

A training record should be kept of all fire safety training given to employees. The following information should be detailed in a training log:

- Date of training, fire drill or evacuation.
- Duration of the training.
- Details of the success of the evacuation, e.g. times, assistance given to disabled people, whether the drill achieved its aim and so forth.
- Name of person providing the training.
- Number of people receiving the training.
- Any comments from employees regarding the effectiveness of the drill or evacuation.

Information and Instruction for Other People

The Management of Health and Safety at Work Regulations 1999 requires employers to ensure that information is passed on to people other than employees regarding the risks to health and any control measures that are required to reduce the risk.

It is essential that people other than employees are given adequate information and, where appropriate, training in the action to take in the event of a fire. In its simplest form, this might be a simple induction at the start of employing contractors. Such induction would make them aware of the risks that exist within the premises, the fire alarm system, the action to take in the event of a fire and the need to keep escape routes, fire exits and so forth clear from debris.

In more complex situations, or where a contract is likely to last for a considerable time, contractors may need to be given more detailed information, instruction and training, including attending fire drills and evacuation.

Training in the Use of Fire-Fighting Equipment

If equipment has been provided for fire-fighting, such as fire extinguishers, fire blankets and so forth, it is essential that employees be given sufficient training in their use. It is perfectly conceivable that a person may pick up a fire extinguisher if a small fire starts near their place of work. Therefore, it would be completely unacceptable if an employer failed to provide basic training in the use of extinguishers. A good rule to remember is that, 'if it is foreseeable, then it will probably happen'. Basically, this means that it is foreseeable that a person will try to extinguish a small fire and, as such, training should be provided as the event is foreseeable.

Training in the use of fire-fighting equipment should include the following:

- Basic instruction on how to operate the various types of fire extinguishers.
- Basic instruction regarding the different types of extinguishers and their uses.
- Instruction to ensure that employees do not tackle fires that are too large for their capability or when they are on their own.
- Instruction in the use of fire blankets for fighting fire and for extinguishing fire in a person's clothing

Summary

Fire risk assessment is an extremely important tool as far as fire safety management is concerned. Assessments can help to identify weaknesses in existing controls and can help identify additional reasonably practicable controls that could be put in place. Control measures come in many different forms–sometimes they are in the form of actual hardware items and sometimes it is more appropriate to control fire hazards by procedures and documentation.

Regardless of the type of control measure chosen, it is important to think of fire risk assessment as not a one-off fix to solving a problem but part of a continuous process. Control measures need to be constantly updated and continually improved upon. Fire risk assessment is the first step in the continuous improvement process.

REFERENCES

HSE Books (1999) *Fire Safety – an employers guide*, HMSO.
HSE Books (1997) *Successful Health and Safety Management*, HMSO.

5 Human Factors in Fire Safety

INTRODUCTION

In May 1985, a major disaster occurred at the Bradford City Football Stadium. Fifty-six people died and hundreds were injured after a fire ripped through the wooden football stand. One of the questions that was posed to the inquiry after the incident was:

> Why had so many people been killed in what was basically an open structure with plenty of room to escape onto the pitch?

Human behaviour, always a factor in such incidents, played a significant role in the deaths of so many people.

The stand, where the fire started, had a capacity for 2000 people, was constructed in 1908 and was made mainly from timber. It ran approximately 90 m long and, as it was constructed on a hill, there was a void underneath the stand, which varied between 225 mm and 750 mm in depth. The stand was divided into two longitudinal sections, each separated by a timber fence approximately 1.5 m in height. Access to the seating area was via a corridor, which ran the entire length of the stand at its highest point.

A council engineer had visited the football stadium in 1984 and had warned the football club that the stand was in poor condition. The club was also told that evacuation should be achievable from the stand within two and a half minutes. Further warnings came from the police, who advised the club that the stand was in a dangerous structural condition. Two days after the fire, work was scheduled to take place on repairs to the stand, funded by a grant from the Sports Grounds Trust.

At 3.00 p.m. on Saturday, 11 May, as the turnstiles were closed and gates locked to keep intruders out of the ground, Bradford City started their last match of the season. At about 3.30 p.m., some of the gates were opened to allow the players' guests to enter the clubhouse.

At 3.40 p.m., shortly before half-time, the fire was discovered. Television coverage of the match broadcast the pictures live as viewers watched the first flames emerge from the stand. These quickly developed over the next five minutes to engulf the entire stand. People desperately tried to escape from the stand, many trying to leave by the routes in which they had entered. As they found the exits barred, some turned back and tried to escape onto the pitch.

When the fire was first discovered, many people within the immediate area of the fire continued to watch the match, as they thought that the fire was a minor one and would be extinguished. However, the majority of the people within the stand knew nothing of the fire and continued to watch the game, completely unaware that a tragic event was rapidly unfolding.

Three minutes after the fire was first noticed, police started to ask spectators to move from the stand. However, many people did not take the situation seriously and remained in their seats. Television footage clearly illustrates that, by 3.44 p.m., flames were clearly visible and spreading quickly. Yet, many people remained in their seats, seemingly unconcerned by the events developing. In fact, some people could be seen actually watching the flames develop without moving from their seats.

Six minutes after the initial fire started, the stand was completely engulfed in flames. Many people by this point had started to spill onto the pitch, scaling brick walls to escape from the intense heat, flames and smoke.

However, most people attempted to escape by the route through which they had entered the football stadium. Although some of the exits had been opened by the police, the majority of exits remained locked. Some of these padlocked exits were forced open by people outside the ground, thus allowing spectators to escape to safety. However, most of the people who died were found near locked exits and turnstiles.

The inquiry into the incident was conducted by Mr. Justice Popplewell, who stated:

> The clearest lesson to be learned is first, that fire can develop at a frightening speed and that the ordinary spectator who has never in his life been exposed to it before will react slowly and second, that the panic and confusion caused by a fire and the smoke that accompanies it is likely to be a source of considerable danger by itself.

Other points that came from the inquiry included controversy between who was ultimately responsible for emergency control of the public and legal implications as to the inspection and control of sports grounds. However, one of the main outcomes from the disaster was the need to study in more depth aspects of human behaviour in such situations. Why had people reacted so slowly to a fire developing nearby and why had people attempted escape from the ground via turnstiles?

Human behaviour in fire situations, regardless of whether the fire occurs in sports grounds, shopping malls, or small workplace premises, can be influenced by the following factors, each of which need to be taken into consideration when carrying out fire risk assessments of the workplace:

- Physiological effects from fire.
- Immediate perception and action when a fire is discovered.
- Motivation to escape from fire.
- Escape routes and escape signs.
- Communication in emergency situations.

PHYSIOLOGICAL EFFECTS FROM FIRE

There are many poisonous fumes produced when products burn but three of the most common gases can prove fatal. The first of these is carbon dioxide gas, which is an asphyxiant in very small amounts and can rapidly affect breathing. In large concentrations, carbon dioxide is a powerful poison.

Carbon monoxide is another gas that is always present in the fumes produced from fires. Carbon monoxide is poisonous in any concentration and affects the bodies mechanism to uptake oxygen. This is very significant in terms of human behaviour because as the body is subjected to carbon monoxide, the cells needed for the thought process get starved of oxygen, resulting in a decreased ability to function properly. A person who breathes carbon monoxide will quickly become confused, with reduced ability to reason properly. They become disorientated and lose complete sense of direction after exposure to the gas, however small it may be.

In addition to nerve cells becoming deprived of oxygen, muscle cells are also affected, making a person less able to physically move away from the fire. Muscle fatigue sets in very quickly after exposure to carbon monoxide, producing a lethargic effect on the person.

A common situation occurs when people wake up in the middle of the night, detect that a fire has started in their house and start to make their escape. However, when they stand up in the smoke, they quickly become disorientated. A simple task such as finding the exit door of their room becomes a very difficult task. The more a person starts to panic as a result of not being able to find his way out of a room, the more his breathing rate increases. This has the effect of taking in more carbon monoxide, resulting in more disorientation. Eventually, after several minutes of panic, the muscles start to tire and the person collapses, unable to move any further. Fire victims are often found lying close to their beds or close to doors. In the workplace, victims are often found near the escape routes or the fire doors.

Cyanide may also be present in the fumes from a fire, depending on the type of material that is burning. If the material contains a high percentage of nitrogen-based substances, cyanide will probably be given off in high concentrations. Very small amounts of cyanide can be fatal in seconds.

Hydrochloric acid is often found in the fumes from fires. This substance has the effect of irritating the eyes and lungs, causing pain and breathing difficulties. The ability to escape is further reduced by lack of visual awareness.

It is important for the fire risk assessor to consider possible situations when people are trying to escape. What might seem an obvious thing for an escaping person to do may well not be the actual action that the person may take. Any help that a person in difficulty can get as they try to escape from a fire can be greatly appreciated.

IMMEDIATE PERCEPTION AND ACTION WHEN A FIRE IS DISCOVERED

Perception is an interesting concept when people are faced with danger. Several in-depth case studies have been carried out on the subject of human behaviour in fire situations. A clear message emanating from these studies is that people tend to be complacent where fire is concerned. This is perhaps because few people are ever faced with a fire situation. The majority of the public, who are also employees in workplace situations, will never experience a fire other than controlled display fires in the open.

Kings Cross Underground Fire

Complacency was demonstrated clearly at the Kings Cross underground station fire in 1987. At approximately 7.30 p.m. on 18 November, reports were made of a fire on the escalator on the Piccadily Line. However, even while travellers identified that visible signs of a fire were obvious, they did not necessarily try to make their way to safety.

At the inquiry, one woman recalled entering the station with other people at about 7.40 p.m. She saw that the ticket hall was 'grey and hazy' and suspected that there must be a fire somewhere. She decided not to use the underground station and made her way to safety through exits onto the street. However, she also noticed that, as she was proceeding out of the station, people were moving in the opposite direction to her. Before leaving the building, she noticed thick black smoke behind her, yet there were still people moving towards the fire.

Woolworth's Fire

Another fire situation that is interesting is the Woolworth's fire in Manchester in 1979. Nine members of the public died in the fire and one member of the staff also died. About 100 people were dining in the restaurant when one of the store managers, who had been alerted of the fire, shouted, 'fire'. However, even though smoke could be seen from the dining area, some customers took no notice of the warnings and continued to eat their meals.

The witness reports from the Woolworth fire and from Kings Cross and Bradford, all seem to indicate that the general public, and even employees of the companies involved, do not necessarily take any notice of fire warnings. People tend to carry on with their normal activities even when they are initially warned that a fire has started. Complacency is perhaps the biggest factor for this behaviour. Unless a person has been the victim of fire in the past, people generally tend to think that it cannot be a serious incident and that someone would deal with it.

People Need more than One Cue

Another interesting outcome of the witness testimonies from these fires and from other fire incidents is that people tend to need more than one cue before they treat the situation as serious. For example, in the Kings Cross fire, travellers did not respond even when they saw the smoke; instead, they preferred to investigate for themselves. This is perhaps another case of complacent behaviour, together with human curiosity. Even the ticket collector at Kings Cross did not respond until he had been told three times that people had suspected a fire. His comments were of the order, 'you're the third person to say that. Perhaps I should go and look!'

Going to take a look is a common factor in most fire cases. People are curious and need to be deterred from proceeding towards a fire. It appears that cues are needed before people will eventually take the message that something is wrong.

Cues can come in many forms. Evidence from the Kings Cross fire suggests that people were immediately willing to take the advice of fire officers or members of the police force. When told by these people to evacuate, travellers did so immediately. However, when advised by members of staff or by other travellers that a fire may be developing, the general public did not take notice and preferred to investigate for themselves.

Model of Human Behaviour in Fires

In the1980s, the University of Surrey developed a model of human behaviour. There are five key aspects to the model proposed by the researchers:

1. *People's behaviour in fire depends on the roles that they perceive to be relevant to their responsibilities in the organization.*

   Generally, the majority of the public tends to react slowly to emergency situations and tends to seek advice from members of staff or from those in authority.

   This situation is confirmed with the Piper Alpha oil fire in 1987, where 167 men lost their lives as a result of an explosion and fire in a North Sea installation. A large number of men, who initially believed that there was a fire, headed for their accommodation module and stayed there, awaiting instructions from their superiors. They actually suffered heat and smoke inhalation while waiting, yet they did not move. However, as it became obvious that no such instructions were to be received, the men took action upon themselves.
2. *People prefer to use familiar routes for escape.*

   Providing special purpose fire escapes is standard practice in most workplaces. However, these might not always achieve their intended aim because most people will not actively seek out an alternative means of escape, but will prefer to use the route by which they entered the building.

   The Bradford City fire illustrated this point, whereby most of the bodies were found near the routes through which people had entered the stadium.

   This is extremely significant when assessing the risks from fire. Escape routes should be regularly incorporated into the normal access and egress from a building. This will not only make people aware of their existence but will give employees an opportunity to actually use them. In some cases, it is advisable to block off normal access routes during fire drills, forcing people to seek out alternative routes of escape.
3. *People tend to disregard alarms.*

   Fire alarms can be confusing and often leads to people ignoring them or reacting slowly to them.

Clear, unambiguous alarms need to be installed in the premises. These should be tested regularly, preferably at times when people are not expecting fire drills. Response times should be recorded and, if found to be too long, action should be taken to seek out methods of increasing responses from employees.

4. *If early clues to fire are ambiguous, people will seek more information for evaluation.*

   It is obvious from studies and from past incidents that, generally, people cannot be expected to react properly to only one signal or cue. Even if people hear a fire alarm, they will tend to want to investigate for themselves. Unfortunately, the delay in trying to gain access to further information might well prove fatal for people.

5. *People do not always react in their best interests when fire is indicated; such behaviour is often interpreted as panic.*

   The fact that people try to gather intelligence after the first warnings that a fire may be present is evidence that they will act on information received. This is an extremely important point as far as planning for an emergency is concerned.

   Safety management and fire safety management should be centred around this fact.

MOTIVATION TO ESCAPE FROM A FIRE

The suggestion that people require motivation to escape from a fire seems to be very unusual. However, research does show that if people are engaged in an activity, such as eating a meal or being entertained by some event, they may be much less willing to evacuate after hearing a warning.

This situation seems even more bizarre when people can actually see a fire developing. Remember the Bradford football stadium incident, where people actually kept watching the match while they could see the fire.

Motivation to escape where people can see a fire seems to be clearly flawed. Security cameras in shopping centres have actually caught on camera people looking through racks of clothing while a small fire develops nearby. One shopper in particular actually turned her head to see the fire, yet remained in the shop and continued to rummage through racks of clothing.

It appears that motivation to escape from a fire is even less when people cannot see the fire or smoke. Instead, they try to investigate further.

FIRE RISK ASSESSMENT AND HUMAN BEHAVIOUR

Considering the aforementioned points about human behaviour, it is perhaps difficult to see how the situation can be improved. As part of the fire risk assessment process, human factors should be considered and questions should be asked as to how likely it is that people will react in a certain manner.

A fundamental aspect of fire safety management is to consider in detail the benefits of training people to escape from fire. Most companies carry out fire drills on a regular basis, which is often a requirement of the fire certificate. This practice should continue and is a good method of continuing the awareness process.

However, most companies, when carrying out fire drills, do so in a similar manner each time. People get accustomed to evacuating the building by a set route–usually the one they have taken to enter the building in the first place.

Fire drills should be carried out whereby the usual route of exit is blocked temporarily, forcing people to use other exits. If different exits are used each time a drill is carried out over the course of a year, employees would perhaps become familiar with all exits, rather than just the ones they normally use.

COMMUNICATION SYSTEMS

A study carried out on behalf of Tyne and Wear Passenger Transport Executive in Newcastle looked at the communication systems that are typically used to alert people of a fire situation.

The study was carried out at a complex commuter station. Different information was supplied to people each time an evacuation had to be made from the station. The information given to users in each situation was as follows:

1. Alarm bell only.
2. Alarm bell with two staff members present to assist.
3. Alarm bell with minimal non-directive public announcements.
4. Alarm bell with two staff members present and directive announcements.
5. Alarm bell with improved directive public announcements.

The study involved assessing three criteria:

- The time to start to move.
- The time to clear the station.
- The appropriate behaviour in the situation.

The results of the study are quite complex but generally reveal that evacuations 4 and 5 were completed rapidly and safely in comparison to the other evacuations. The study highlighted the importance of issuing prompt instructions to the public and employees, explaining exactly what is happening and providing clear instructions as to what to do to safely evacuate the premises or building. It is also clear from the study that people, who do not get such clear information, will delay and even try to investigate what has caused the alarm.

SUMMARY

It is very clear from various fire incidents and from research that employees and the general public do not often behave in the manner in which we may anticipate them to. People will delay evacuating, particularly if they are engaged in work that they think is important. Even once they do decide to respond to a fire alarm, they will often try to gather more information rather than immediately evacuating the building.

The role of the fire risk assessor is to be aware of these aspects of human behaviour in fire situations and to develop ways of foreseeing what is likely to occur. The following should be taken into consideration when planning for fire safety:

- Clear unambiguous fire alarms are required to communicate that a fire or other emergency has occurred.
- Fire alarms on their own will not necessarily prompt a person to evacuate from a building.
- Re-enforcement of the initial fire alarm may be required and should be provided as early as possible. This can take the form of public address (PA) announcements giving clear instructions or visual display announcements.
- Someone needs to take immediate responsibility for evacuation. The designated person should be clearly identified and known to members of staff.
- Fire wardens may be appropriate in certain situations. Wardens can be very helpful in ensuring that the premises are emptied and, if trained properly, can provide a great source of information to people while trying to escape from a building.
- Fire drills should be conducted regularly and should involve the use of alternative means of escape.

REFERENCES

Fire Engineers Journal (1994) *Action in the event of fire: Human behaviour – a firefighters view*, issue number June 1994.

BRE Digest (1993) *Human behaviour in fire*, Building Research Establishment.

Fire Prevention Magazine (1985) *The Popplewell Inquiry*, issue number 183.

Fire and Materials (1999) *Fire alarm in a public building: How do people evaluate information and choose an evacuation exit*, issue number 23.

6 Arson

INTRODUCTION

On 25 September, 1982, a fire and explosion occurred at a warehouse in Salford, causing extensive damage to the warehouse, loss of business and widespread minor damage to surrounding buildings and residential property. More than 2000 tonnes of chemicals, including 25 tonnes of sodium chlorate, were stored within the premises, which, after ignition, produced a violently explosive mixture.

The main activities of the company were storing, packing and transporting chemicals. A significant part of the process involved the use of sodium chlorate, which is principally used as a weedkiller. Under fire conditions, it decomposes at approximately 265 °C, liberating oxygen and heat in the process. When sodium chlorate is mixed with other organic materials, the rate of burning is significantly increased.

The warehouse occupied part of an industrial site that was undergoing major redevelopment work. Owing to a lot of demolition work, there had been serious problems with vandalism throughout the industrial site. During the 20 months leading up to the fire and explosion, about half of the fires in the area had been started deliberately in unoccupied premises.

It was reported that one night, children had gained entry into the company's chemical warehouse yard and had vandalized part of a load on a lorry. Nine days before the fire at the warehouse, vandals had started a fire against one of the walls of the warehouse, using rubber tyres as a fuel source.

At approximately 11.20 p.m. on Saturday, 25 September, the night watchman, who was employed by a contractor security company, left the gatehouse in the yard to carry out his routine hourly inspection. He noticed flames coming from a hole in the roof of a part of the warehouse. Just over a minute later, a violent explosion occurred, causing major structural damage to the warehouse. Many flats and houses in the area suffered blast damage, especially to the windows and roofs.

More than 60 persons were treated in hospital, the majority suffering from inhalation of smoke and toxic fumes and from shock from the explosion.

Earlier in the evening, the police received reports of vandalism in that area. It was later suggested that it was probable that a fire had been started externally between 10.30 p.m. and 11.00 p.m. It was likely that this fire had spread quickly to the roof of the warehouse. One eyewitness noticed 'rolling' flames around the roof, which is indicative of a highly flammable mixture of gases, an explosion being imminent.

The fire in the roof and the fact that a flammable atmosphere was present would have generated enough heat to start decomposition of the chemicals stored in the warehouse. Once decomposition occurs, it can take fractions of a second to cause a violent explosion.

Following police investigations into the fire and explosion, five persons were charged, with two of them facing charges of arson.

In conclusion, the Health Safety Executive's (HSE) investigation into the fire and explosion stated that:

> Although storage arrangements for sodium chlorate and other chemicals could have been better, it was likely that vandalism played a major part in the sequence of events.

One of the recommendations made as a result of the investigation was that employers should consider reasonably foreseeable events from vandalism and from arson attacks and should take suitable measures to eliminate or control risks to protect their premises.

In the above case, it was perfectly foreseeable that an arson attack could take place. Evidence of large numbers of fires in the surrounding areas was available and there had even been past cases of fires started deliberately within the yard of the warehouse. This is one of the most common signs that a high risk of arson is present. These signs should prompt employers to start thinking seriously about possible arson attacks.

A similar situation occurred more recently at an MFI store in Gloucester, where a fire was started deliberately in the yard, causing £5 million of damage. The fire was the latest in a series of fires in and around the site. People in that area had seen young lads setting fire to a skip several days before the MFI fire.

STATISTICS RELATING TO ARSON ATTACKS

Current statistics show that 22% of fires are started deliberately within residential premises. However, 50% of fires in industrial, commercial and public premises are due to arson attacks.

Almost 100 people are killed every year as a result of fires started deliberately, with more than 3000 people being seriously injured.

Insurance companies in the United Kingdom pay out more than £1 million per day in compensation claims as a result of fires in the workplace. Many people assume that such compensations will fully compensate for loss; however, this is not the case. Insurance claims will not cover items such as computer records of customers; records of employees and contracts; vital sales information; and procedures for carrying out specific processes.

Arson is often thought of as an inconvenient fire but many believe that the losses incurred as a result of an arson attack will be recovered through insurance. However, losses from arson attacks are not purely financial, they can also involve the following:

- Loss of life of employees, visitors, emergency crews and even members of the public.
- Injury to any of the above groups of people.
- Business interruption and, increasingly, complete bankruptcy of the business.

- Job losses.
- Loss of economy to the local area.
- Pollution of the atmosphere from smoke, fumes and fire-fighting run-off waters.
- Loss of historical premises or loss of art collections.

Arson fires tend to cause much more damage than accidental fires because they are often started with multiple points of ignition and may also be assisted with flammable materials such as petrol. A person starting a fire deliberately will probably choose vulnerable points in a building, which have the effect of weakening the structures and causing the fire to spread rapidly.

Although it is possible to insure premises against arson, human suffering and business interruption may cause substantially more in terms of loss from arson. Between 1990 and 2000, more than £5700 million worth of claims have been submitted as a result of malicious fires. However, uninsured loss in terms of business interruption is also very significant. The Association of British Insurers (ABI) estimates that £1500 million in business interruption was also caused because of arson attacks during that period.

Over the last ten years, an average of 2500 individuals have been charged or cautioned for arson attacks. In 1997, 63% of those taken to court were found guilty. Of these, the following sentences were imposed:

- 227 were placed on probation
- 163 were fined
- 361 offenders were made subject to a supervision order or community service
- 20 were given a suspended sentence
- 204 were given a conditional discharge
- 5 were given an absolute discharge
- 458 were given a custodial sentence.

Arsonists range in ages from 10 upwards. A significant number of offences are carried out by individuals in the age group of 10 to 17, suggesting that the majority of arson attacks are not premeditated, instead they may be opportunistic if the right circumstances occur. This is significant in terms of controlling the risk due to arson.

Arson attacks are prevalent in every industry sector; however, the largest percentage of fires started deliberately belong to the recreational or cultural category, which accounts for almost 66% of the fires in this category. Surprisingly, the lowest percentage of fires started deliberately is in the domestic dwelling category. The following table lists some common industry sectors and their risk in terms of arson attack. Those at the top of the table have the largest percentage of arson attacks in relation to their overall fires.

1. Recreational and cultural
2. Education
3. Construction
4. Agricultural premises

5. Retail distribution
6. Hospitals
7. Hotels and catering premises
8. Other industrial premises.

SIGNS OF ARSON ATTACKS

It is often difficult to determine whether a fire started in a building is due to an arson attack or is purely accidental. However, there are some signs that can possibly indicate arson rather than accidental ignition.

Fires in a Number of Different Sites Within Premises

One of the most obvious indicators of an arson attack is that a number of fires have started around the premises or within a building. An arsonist will often try to make sure that the fire gets a good hold before he withdraws from the building. To do this, fires are started at various points.

Use of Accelerants

Arsonists often use accelerants such as petrol or solvents to ensure a rapid spread of fire. Residues from these materials can sometimes be detected by forensic scientists.

Fire Doors Wedged Open

A good indicator of an arson attack is that fire doors have been left open throughout the building. These are sometimes wedged open; however, in many cases, fire extinguishers are often used to wedge open the doors. Left open, fire doors will greatly assist the fire and smoke spread and will quickly contribute to the fire spreading throughout a building.

Fire System Disabled

Serious arsonists will often disable fire-detection and fire-fighting systems to ensure that fires starting in a building will spread quickly before they are detected.

A MOTIVE FOR ARSON

Although the majority of arson cases are due to bored youths with no particular grudge against their victims, there may be some other motives for deliberately setting fire to the workplace. The following are some of the common motives for arson attacks.

An Employee with a Grudge Against the Company

Several cases of arson every year are due to employees with some type of grudge. It might be that a person has been made redundant and feels unfairly treated. Given the opportunity and the right circumstances, such a person may decide to repay their employer with an act of vandalism.

One case that was highlighted in court involved an employee who had invented a new process but the employer had failed to recognize his important work. In order to express his important role in developing the process, the employee decided to 'get even' with the employer and started a small fire, which was aimed at causing a small amount of inconvenience. However, the fire grew rapidly and destroyed an entire laboratory.

This type of attack is rarely carried out after working hours. Past employees who decide to take revenge often try to gain access to the company's building during working hours, possibly relying on their past contacts to help them gain access. This type of attack should be relatively easy to prevent. Good security arrangements such as safeguarding keys and other access devices is often all that is necessary to prevent such attacks.

Employees Who have Committed Fraud and Hope to Cover their Tracks

Although rare, it is possible that fraudulent employees may wish to cover their acts by destroying corrupting evidence, especially computer files. One of the most effective ways of carrying out such action is to start a fire near the computer equipment. A small amount of heat will start decomposing plastics on computers and will then spread to the actual electronic components, causing irretrievable damage.

To Hide Another Crime

One of the most common reasons for an arson attack is that burglars or other intruders try to cover their crime by destroying evidence of forced entry.

Those who break into premises to steal money, equipment or other items often try to hide their crime. The majority of such attacks are often opportunistic as well. For example, a person who breaks into premises usually does so without taking fire-raising equipment with them. Instead, once they have committed their first crime, they may look for ways of concealing evidence, and may choose to start a fire as a possible option. Any combustible material left lying around will provide the means of carrying out their second crime. Suitable combustible material might include waste rubbish, chemicals left lying around and other flammable material.

A recent fire in a retail premises was initially put down to a fire starting in the electrical system. The point of origin clearly indicated that the fire started near an electric socket. It was only when the safe was discovered missing that arson was considered a possibility!

Opportunism

Although there are many motives for arson, the majority of arson attacks are carried out by opportunists, mainly young persons who have taken the opportunity, if it presents itself, to cause damage. Skips lying full of rubbish and left next to a building present a great opportunity for these people. Setting a small fire in a skip would be relatively easy; the fire may then spread rapidly to an adjoining building.

Another situation in which an opportunity presents itself is where windows have been left open. Perhaps burglary is not on the minds of bored youths; however, malicious damage may be. It is again relatively easy for these people to throw burning rags or other lit material through windows, and these materials can then spread fire inside a building.

Therefore, in most cases, there is no clear motive for arson. The ironic thing is that this type of attack is probably the easiest to prevent. Effective security measures, both inside and outside buildings, goes a long way towards preventing opportunist-type arson attacks.

Pyromania

Finally, pyromania is another motive for arson. This is where the arsonist suffers from an illness and starts fires as part of that condition. A common type of compulsive pyromania is seen when the individual starts a fire, usually in domestic premises, and then attempts to save the victims. A number of people have been convicted of such crimes, often after they have been hailed as heroes because of their efforts to rescue trapped fire victims.

FIRE RISK ASSESSMENT

As part of the overall approach to fire risk assessment, arson should be considered as one of the hazards. Like all other hazards, this must be fully evaluated for risk. The checklist presented in Table 6.1 can be used to identify and control a possible risk from arson.

CONTROLLING THE RISK FROM ARSON

After carrying out a fire risk assessment to identify the risk from an arson attack, suitable control measures should be put in place. Control measures can be broadly divided into the following categories:

- Security
- Passive fire protection
- Active fire protection
- Housekeeping
- Contractor's policy
- General lighting.

Table 6.1

Vulnerability of the building – internal.	Are there any vulnerable points within the building where fire could be started, e.g. unoccupied areas, areas where there are no security alarms, areas where there are no forms of fire detection?
Vulnerability of the building – external.	Are there areas outside the premises where fires could be started deliberately, e.g. skips with combustible materials inside, rubbish accumulating around the site? Is there a form of site security? If there is no site security in place, are there areas around the site where the lighting is poor? Good general lighting is one of the best deterrents against an arson attack.
Location of the premises.	Is the company based in a location known for its crime rate, especially crimes involving those of an opportunistic nature?
Combustible materials.	Are there any likely combustible materials, both inside and outside the building, e.g. solvents, petrol and other flammable materials? Are these stored securely after working hours? Are vehicles stored securely, e.g. fork-lift trucks?
Persons who could start fires.	Identify which categories of persons could start fires deliberately, e.g. employees, past employees, intruders or visitors.
Security.	Are keys and other access devices reduced to as few a number as possible? Are keys and other access devices taken from employees who leave the company? Are keys and access devices logged and accounted for? Where the above does not occur, are locks and combinations of locks changed regularly? Are checks carried out at the end of working hours to ensure that the premises are secure, e.g. windows closed, doors locked, fire doors not wedged open? Does a person have responsibility for ensuring that the premises are secure (if not, it will not get done!)? Is external lighting adequate? Are there areas around the site where the lighting is poor? Is there a security system, e.g. closed circuit television or an alarm system?
Assess the risk.	Has the level of risk been determined?
Emergency plan.	Has a business recovery plan been written into the fire emergency plan?
Review the risk assessment.	Has the fire risk assessment and the arson assessment been reviewed regularly?

Security

Security measures are probably the most cost-effective control measures that can be put in place to prevent an arson attack. Effective security arrangements will certainly deter the opportunist-type attack and will go a long way towards deterring other types of arson attack.

As previously discussed, steps should be taken to ensure that buildings are secured at the end of working hours. This should include buildings used for storage, such as sheds, garages and portable structures. A single person should be nominated to ensure that premises are secured at the end of each working day. Keys for each individual building should be locked securely in appropriate cabinets and access should be strictly limited.

Securing the perimeter of the site should also be considered. Although fencing might not keep a determined person out of the site, it will certainly delay them, perhaps long enough for them to be detected. Where the perimeter is shared by several employers, it is often worth meeting with neighbours of adjoining commercial properties to reach an arrangement whereby perimeter fencing could be constructed and the costs can be met jointly.

Existing fencing should be checked regularly to ensure that it maintains its integrity and any maintenance work required should be carried out promptly.

Full-time security services may be expensive to employ; however, it really depends on the nature of risk whether or not these should be considered. If the risk from arson is high, e.g. in an area where young persons are known to start fires, then the cost of employing 24-hour security might be the only solution that offers an effective deterrent.

Another aspect of security that is often overlooked is that of employees gaining access to premises. In most cases, employees can be fully trusted to carry keys and other access devices such as swipe cards or can be given combination lock codes. However, employees may pass on such information, perhaps by accident, to others in conversation. It is important that a detailed account is made of all the keys to the premises and all non-essential persons limited to access. The following questions could form part of the fire risk assessment as far as security is concerned:

- How many employees have access to keys and other security information?
- How many people actually need such access?
- Is there a method of determining who can freely enter the premises outside normal working hours?
- Are employees who leave the company asked to hand in their keys?
- Are security codes changed when someone leaves the company?
- Are security combination codes changed at frequent intervals?

The last point is an important one. Most computer users have to change their password information on a regular basis. This is to stop information about access codes getting to unwanted persons. However, combination codes for doors, padlocks and other devices are rarely changed, even when people leave. Poor management of security may lead to a successful arson attack!

Passive Fire Protection

The purpose of passive fire protection is to contain any fire that does break out within a building. Passive fire protection includes compartmentation, fire doors, protected lobbies and stairwells.

These structures should not be altered in any way without first contacting the local fire authority. It is important not only for arson attacks but also for general fire protection purposes that these structures are checked regularly to ensure that their integrity is maintained. Fire-resisting walls and compartments should never be breached by pipes, cables or other wiring, without fire-stopping material being installed.

Fire doors should not be wedged open during normal working hours. It is also important that checks are carried out at the end of each working day to ensure that fire doors are closed properly. It is bad practice to allow fire doors to be occasionally wedged open, as this leads to complacency.

One important consideration as far as arson is concerned is that fires may be started deliberately by throwing material in from outside a building through windows, letter boxes and any other gaps that might occur. A number of factories might have been the subject of significant damage because they had small gaps in roofs, which allowed people from the outside to throw lit material into the building via these gaps.

Letter boxes are one of the most common places for arsonists to strike. Lit material is easily forced through letter boxes and it will then spread flames to other combustible material inside a building. A simple solution is to fit metal catch bins for mail and for other unwanted material. These prevent lit material from dropping onto the floor and spreading flames to the floor and the wall coverings.

Active Fire Protection

Automatic fire-detection panels should be situated near the main door or the entrance. These panels provide the fire brigade with important information, which may lead to rapid identification of the area where the fire is located and hence to speedy action to extinguish the fire.

Active fire protection systems are designed to fight any fire that has broken out. These include fire detection, which may be automatic or manual, and fire-fighting equipment such as sprinklers.

Although rare, it has been known for arsonists to sabotage valves of sprinkler systems in an attempt to stop active fire systems from working properly. Rooms or areas containing sprinkler valves should be protected from intruders by suitable protective systems and should be inspected regularly to ensure that they are functioning properly.

The following precautions can help deter attempts to sabotage active fire protection systems:

- Valves should be located together in a single room for ease of protection.
- Valves should be clearly identified and should be strapped and padlocked in the open position. The open position should be labelled.
- The local fire brigade should be informed about the position of valves and their operation.
- Any fuel intakes for pumps, engine cooling and exhaust outlets for equipment used as part of an active fire protection system should be located in areas to ensure that they are not tampered with.

Housekeeping

One of the most common types of arson attack is opportunistic in nature, especially where young persons are concerned. Bored youths, who for some reason decide that it may be entertaining to set fire to property, are one of the largest groups of arsonists.

Opportunists do not generally go to great efforts to break into premises to start fires. Instead, they rely on materials being stored externally to buildings, and these materials may then be used to spread fire into the buildings. Examples of the types of materials used to start such fires include the following:

- Skips overfilled with combustible waste.
- Refuge bins left out, awaiting collection.
- Combustible waste left lying near walls of buildings.
- Discarded solvent and petrol containers left lying in yards.

It is important to recognize that most fires started deliberately will involve one of the above. Rubbish should not be left in open unsecured yard areas. Instead, skips, bins and other such collection vessels should be stored internally or in areas where they can be secured from tampering. Where these have to be stored in yards, they should be left well away from the walls of the buildings and should be positioned in areas that are well lit and preferably in areas where there are people going around.

Contractor's Policy

Another important area for control is the contractor. It is pointless going to the trouble of putting in place security arrangements if these are going to be rendered useless by contractors.

Before contractors are allowed access to premises, they should be asked to submit detailed information regarding their approach to safety. They should be asked to supply a copy of their health and safety policy and any risk assessments that may be necessary. In addition, if contractors are to be left unsupervised in premises, especially outside normal working hours, they should be asked to supply a policy statement regarding employee security.

It is not enough to purely rely on policies and risk assessments from contractors. Anyone working within the premises should be informed about the fire risks and action to take if a fire is detected or is started accidentally.

Finally, security arrangements need to be passed onto contractors when they are required to lock up premises at the end of their work. Arrangements should be clearly detailed, documented and passed to a person from the contracting company who has a high level of responsibility.

General Lighting

One of the most effective deterrents against arson attacks is effective external and internal lighting. The majority of retail premises on busy streets have some form of internal lighting permanently left on.

In an industrial estate, it is perhaps not essential for internal lighting to be left on. However, good quality external lighting is vital. Care should be taken when carrying out risk assessments to identify any external areas that may be poorly lit and that might allow persons to hide undetected. Areas around main doors or vulnerable access points should be clearly lit. In addition, areas that present a high fire risk also require effective lighting.

These include areas such as:

- storage of combustible waste, e.g. skips and bins,
- stores for vehicles, where there may be flammable fuels,
- flammable liquid stores,
- gas cylinder storage areas,
- main access to premises,
- raw material storage.

SURVIVING ARSON ATTACKS

Statistics show that 80% of businesses that have suffered an arson attack go bankrupt within five years following the attack. This is due mainly to the fact that few businesses plan for arson. Part of the planning process, and the fire risk assessment process, is to develop contingency plans should arson occur. Contingency plans help a business to recover after a major disaster.

Contingency plans should consist of the following:

- Procedures for dealing with disaster, e.g. fires caused by arson.
- Contingency for ensuring customer orders are met on time or that services are provided to clients.
- Details of agreements made with other companies, e.g. companies who supply products or materials.
- Emergency arrangements, e.g. getting help from the local authority, police and others.
- Employee relations, e.g. minimizing disruption to employees.
- Telephone numbers of important contacts.

It is worth remembering that emergency plans may be destroyed in a fire. A copy of the plan should be held in a place away from the business premises, e.g. at home or in a bank safe.

SUMMARY

Depending on the type of business or location of the business premises, arson may present a significant risk from fire. In some cases, arson will be the largest risk to business. Although significant damage can result following an arson attack, it is possible to control the problem. By carrying out a systematic risk assessment or by incorporating the risk of arson within the main fire risk assessment, vulnerable areas can be identified. Once this is done, suitable control measures can be put in place to control the risk.

REFERENCES

HSE Books (1983) *The Fire and Explosion at B and R Hauliers, Salford*, HMSO.
Arson Prevention Bureau (2000) *Prevention and Control of Arson in Industrial and Commercial Premises.*

7 Common Causes of Workplace Fires

INTRODUCTION

Although arson accounts for more than half of the major fires in industry, with an average cost of approximately £400 000 per fire, there are many other causes of fires in the workplace. The following are some of the main causes:

- Electrical equipment
- Machinery and plant
- Smoker's materials and matches
- Maintenance work
- Blowlamps, welding, cutting and other hot work
- Heating appliances
- Cooking appliances
- Rubbish and other debris.

ELECTRICAL EQUIPMENT

The main causes of fires in electrical equipment are:

- general overheating of cables caused by overloading of circuits,
- damaged electrical insulation on cables,
- damaged portable equipment,
- local overheating caused by poor connections, which may cause sparks that can ignite combustible material,
- combustible material left too close to equipment which may ignite owing to the heat given off from the equipment,
- ventilation failure caused by rags or other similar materials being placed across ventilation points of electrical equipments,
- wrong fuse fitted.

There are generally two categories of fires involving electrical equipment. Firstly, there are fires involving the actual electrical wiring system; secondly, there are fires involving portable electrical equipment. Fires in the first category are rare, mainly because most workplaces have modern wiring systems. The majority of fires in the workplace that are attributable to electrical equipment come under the second category, which involves the use of portable equipment. Generally, portable equipment is anything that operates from the main electrical supply, connected via a wired plug.

The main legislation relating to electrical equipment is the Electricity at Work Regulations 1989, which requires that all electrical systems should be:

- constructed and maintained to a standard that prevents danger,
- regularly checked and tested by a competent person,
- suitable and sufficient for the environment in which it is to be used,

- protected by earthing, double insulation and/or RCDs,
- capable of being made dead (isolated),
- use of low-voltage equipment whenever possible, especially when used outside.

Portable electrical equipment should be assessed for fire risk and suitable control measures should be put in place to reduce the risk. Risks can be managed and controlled by setting up an appropriate maintenance system. This should include the following:

- Identification of equipment that has to be maintained.
- Discouraging unauthorized and untested equipment in the workplace, i.e. by not allowing employees to take into their workplace electrical equipment from home.
- Carrying out simple user checks to check for signs of damage, e.g. casing, plug pins, the condition of cables and so forth.
- Formal visual inspection of all items of portable electrical equipment on a regular basis.
- Periodic testing of portable equipment, which should be carried out by a competent person.
- Ensuring that a system is in place for reporting and replacing defective or damaged equipment.
- Training on the safe use of equipment.

Inspection and Maintenance Strategies

There are really three different parts to a strategy of inspection and maintenance:

- User checks.
- Visual checks by a competent person.
- Combined visual and testing checks by a competent person.

User Checks

A system should be developed whereby users of electrical equipment check the item for the following:

- Scuffing damage to cables, especially cables of equipment used outside or in particularly rough environments.
- Damage to plugs and pins, e.g. signs of cracking.
- Joints in cables are strong and secure, e.g. no taped or loose connections where cables have been joined.
- The outer sheath of the cable is secured where it enters the plug and where it enters the appliance or the transformer.
- The equipment is not subjected to adverse environmental conditions without adequate protection, e.g. water, rain and so forth.
- There is no visible damage to the outer casing of the item.

Formal Visual Checks

A unique number should be used to identify each item of portable electrical equipment. A system should be developed to ensure that regular formal inspections are carried out on each item. A suitable logbook should be kept, identifying each item, the visual inspection date and any fault found with the equipment. This ensures that a history is developed for each piece of equipment, thus helping in identifying recurring problems.

The time between regular visual inspections will vary depending on the use and location of the equipment. For example, it may be sufficient to carry out a visual inspection of a photocopier in an office on a two-yearly basis, or even longer. However, a portable drill, which operates from a transformer, used on a construction site may have to be inspected much more frequently.

Combined Visual Inspection and Testing

Occasions when inspection and testing may be required are:

- whenever there is reason to suppose that the equipment may be defective,
- after any repair, modification or adaptation of electrical equipment,
- the manner and frequency of use of the equipment.

The inspection carried out in conjunction with the testing should include the following:

- Checking of the polarity.
- Checking of the fuse requirements.
- Checking of the termination of cables.
- Checking of the suitability of the equipment for the environment.

MACHINERY AND PLANT

Although fires in machinery are relatively uncommon, fires do occur, usually as a result of poor maintenance systems. Plant and equipment that is not properly maintained can cause fires when the following circumstances occur:

- Poor housekeeping, such as allowing ventilation points to become clogged with dust or other material.
- Leaking valves and flanges, which might contribute to oil spills in close contact with hot surfaces.
- Static sparks from poorly maintained surfaces.
- Frictional heat from bearings and loose drive belts.

The Provision and Use of Work Equipment Regulations (PUWER), first introduced in 1992, applies to items of work equipment, from simple hand tools to automatic manufacturing machines. PUWER has many requirements; however, one of the most significant in terms of preventing fires is that employers are required to put in place regular maintenance strategies. These have to be designed and implemented as a result of risk assessment. For example, if a machine presents a significant risk of fire during its normal operation, very frequent maintenance would be required on the machine. In addition, there is a requirement to regularly inspect equipment and to keep a history of the findings and of any repairs that are required.

SMOKERS' MATERIALS AND MATCHES

Smoking in the workplace is a very common cause of fires. Attempting to ban smoking completely will do little in terms of controlling the problem. Human nature suggests that people who have been banned from smoking will usually try to find 'hidden' places to smoke, which presents a greater risk.

It is better to provide specific places for smoking rather than attempt a total ban. This will certainly help unauthorized smoking and, if proper places are identified and proper facilities provided, the risk of fire should reduce.

Metal bins should be standard for smoking materials to be disposed in. Ashtrays and bins should be emptied regularly. Serious fires have occurred in the past where metal bins used for discarding smokers' materials were also used for paper waste. Bins or other receptacles for smokers' materials should only be used for this purpose.

MAINTENANCE WORK

After arson, the most common cause of fires in workplaces is that associated with maintenance and building work. Many serious fires have occurred as a result of poor control over contractors and their work practices. However, it is not only the contractors who are responsible for a high incidence of fires in this category; internal maintenance staff are equally at risk from fires during even the most routine of maintenance tasks.

It is essential that maintenance tasks are properly assessed for their fire risk. Often, a separate assessment is needed each time a building or maintenance task is carried out, unless the task involves a routine operation, in which case it is appropriate to assess it once only.

Fire risk assessments carried out on this type of operation should consider the following:

- Accumulations of combustible waste and building materials in escape routes.
- Accumulations of combustible waste and building materials stored against external walls.
- Accumulations of waste stored externally that might be an easy target for the arsonist.
- The temporary loss of fire exits because of construction work.
- Fire doors propped or wedged open.
- Openings created in fire-resisting structures, e.g. pipe and electric cable runs.

- The introduction of more people who may need to escape in the event of a fire.
- The introduction of electrical equipment, especially portable tools.
- The temporary removal of detection equipment and fixed fire-fighting equipment, e.g. sprinklers.
- Site security.

All the above points should be carefully considered as part of a fire risk assessment that is carried out before building or maintenance work commences. Additional controls should be implemented to reduce the risk during maintenance work. For example, part of a building project may involve the need to remove or cover a fire detector, possibly because a large amount of dust may be produced from sanding. In this case, an additional control may have to be considered during the time the detector is out of action. Examples of additional controls might include the following:

- Regular patrols of the area.
- Withdrawing any other work that may be carried out in that particular area.
- Limiting the number of people who are allowed into that area.
- Additional fire extinguishers.
- Briefing contractors or maintenance staff to ensure that they are fully aware of the fire risks.
- A temporary change to the fire warning system.

In addition to the control measures that may have to be put in place during building and maintenance work, contractors who are employed to carry out such work should be fully investigated before they are commissioned. A company employing contractors should have a contractor's policy, which includes the provision for assessing the competence of the contracting staff. Such assessment should include the following:

- Asking for a copy of their safety policy.
- Asking for details of any accidents or incidents that they have had in the past.
- Asking for details of any Health Safety Executive (HSE) or Local Authority health and safety enforcement action, e.g. fines or improvement notices.
- Asking for copies of their fire risk assessments.
- Asking for copies of their training records.

At present, there are a number of training schemes that are designed for establishing the competence of contractors. One such scheme is the Safety Passport programme. Some large companies and organizations insist that any contractor they employ must have passed this two-day course and must have been awarded a 'Safety Passport'. This provides a good measure of at least some form of health and safety awareness training, which includes a module on fire safety.

However, even if contractors have demonstrated their competence and have displayed their policies, it is still the employer's responsibility to pass on any information about risks within their premises. Contractors should be treated in the same way as employees as far as risks are concerned. For example, employees generally receive regular fire drill training, which is aimed at informing them of the fire alarm, escape routes and assembly points. However, it is unfortunately the case all too often that although employees are given this training, contractors employed in a company are forgotten about. These people also need to be told and made aware of the procedures to follow.

BLOWLAMPS, WELDING AND CUTTING

Any form of hot work can present a significant risk in terms of fire, especially where this type of activity is carried out near flammable or combustible materials. Once again, such work should be fully assessed before it is authorized to proceed. One of the most important forms of control for this type of activity is the permit-to-work system.

A permit-to-work is appropriate in situations of high risk, such as hot work near flammable or combustible materials. The use of permits will ensure that:

- there is a formal check that confirms that a safe system of work is being followed,
- coordination is carried out with other people who may be working in the area, or whose activities might be affected by such hot work,
- appropriate supervision is being provided for people carrying out the hot work,
- provide time limits when it is safe to carry out the work,
- provide additional controls, which may be documented in the permit, e.g. barriers, fire-fighting equipment and personal protective equipment,
- additional permits for specialized work have to be raised, e.g. electrical isolation permits, gas monitoring permits,
- communication between contractors and the commissioning company is being carried out,
- any fire-detection equipment or fixed fire-fighting equipment that has been taken out of service while the work is carried out is reinstated after the work has been completed.

A management system, such as a fire permit policy, ensures that the employees and the contractors know that hot work should not commence until the necessary permits have been raised and authorized. Instructions may have to be sent to contractors, together with a copy of relevant policies to ensure that they are aware of the rules. Permits are a very effective means of controlling high risk situations but they rely on good communication and good control over their use.

As a minimum, permits for hot work should include the following:

- Measures to make sure that all flammable materials have been removed from the work area, or if they cannot be removed, to ensure that they are adequately protected from sparks or heat.

- Provision for fire-fighting equipment to be made available in the work area.
- The permitted time span of the activity and the level of supervision required, which will depend on the nature of risk and the number of people involved in the task.
- The actions to be taken when the work is finished, including initial and subsequent checks to ensure that there are no smoldering or hot materials that could allow a fire to break out at a later time.

HEATING APPLIANCES

Fires involving heating equipment are usually caused by:

- careless refilling of heaters, especially those that use liquid fuel such as paraffin,
- placing combustible materials near, or on top of heaters,
- placing portable heaters too close to combustible materials,
- failure to follow instructions for changing gas cylinders on heating equipment,
- failure to carry out regular inspection of the heaters and associated equipment, especially fuel lines and tubes on portable gas heaters.

Heating equipment should be serviced regularly by a competent person. In addition, employees should be instructed to carry out daily visual checks of portable heating equipment.

Whenever possible, the continued use of liquid fuel heaters should be avoided. If it is necessary to continually use portable heating equipment, a better solution is to install fixed convection heating apparatus. Also, heaters fuelled by gas cylinders should be avoided above or below ground level, as this presents a significantly higher fire risk in terms of evacuation from the premises.

COOKING EQUIPMENT

Fires involving cooking equipment in the workplace are rare, unless the workplace includes residential areas, such as halls of residence. Unfortunately, in such places fires are still relatively common, usually involving chip pans and grill pans.

The obvious control for such fires is automatic or enclosed fryers, similar to the type used in most homes. These units present much less risk from fire and should be an automatic alternative to the older style chip pan.

Other fires related to cooking include the following:

- Grill pans filled with fat and cooking oil, which have been left unattended.
- Matches used for igniting gas appliances dropped onto combustible material.
- Decorative items in a poor state of repair, ignited from flames on cookers, e.g. wall paper hanging off walls.

RUBBISH AND OTHER DEBRIS

It is important that all waste material is regularly cleared from premises and uplifted. Some of the problems associated with accumulations of combustible waste include the following:

- Waste blocking escape routes, stairs, fire doors and so forth.
- Combustible waste being stored near heat sources.
- Waste being allowed to accumulate because of processes, e.g. wood working machinery that generates sawdust and other wood cuttings.
- Waste material being stored near external walls of buildings; this might spread a fire from the waste material to the building.
- Waste stored externally, which might prove an ideal target for arsonists.
- Packaging materials accumulating in the workplace.
- Skips overfilled before being emptied.
- Poor positioning of skips and other debris bins, which are stored too close to buildings or other fire risks.
- Accumulations of waste material or combustible material in areas that are not normally occupied, e.g. basements, roof spaces and so forth.
- Waste material near poorly maintained machinery, which may be ignited by sparks.
- Waste material stored for long periods in warm environments, which may give rise to spontaneous combustion.
- Incompatible waste materials stored together, which may cause an exothermic reaction.

REFERENCES

HSE Books (1999) *Fire Safety–an employers guide*, HMSO.

8 Combustion and Flammability

INTRODUCTION

Fire is a chemical reaction, often referred to as a combustion reaction that produces heat, light and smoke. Although one of the hazards of fire is that it causes burns, the more significant hazard is probably smoke and toxic fumes, which are produced when a material burns.

The reason that the majority of workplace fires produce smoke and toxic fumes is that incomplete combustion takes place. If a fire were to be started in a laboratory using, for example, methane as a fuel, the reaction could be controlled to produce very little smoke. In this case, the reaction would be as follows:

$$CH_4\text{(methane)} + O_2\text{(oxygen)} = CO_2\text{(carbon dioxide)} + H_2O\text{(water)} + \text{heat} + \text{light}$$

The above reaction in the laboratory produces very little in the way of smoke, and although carbon dioxide can be harmful to health, there is little in the way of toxic fumes that are produced.

However, in the workplace, complete combustion does not occur, as the quantities of oxygen available cannot be controlled properly as in the laboratory. As a result, what occurs is incomplete combustion:

$$CH_4 + O_2 = CO_2 + CO + H_2O + C + \text{heat} + \text{light} + \text{other gases}$$

As a result of incomplete combustion, carbon monoxide, which is very harmful to health, is produced. In addition, solid carbon products may be produced, which in effect is the smoke that we see from fire.

The above situation is worsened when the fire involves materials that contain nitrogen, chlorine or other substances. For example, many plastics contain these substances and, when they burn, produce highly toxic fumes including hydrogen cyanide, hydrofluoric acid and sulphur dioxide. These substances, together with carbon monoxide, produce a potentially lethal combination.

The effect that the combination of the above gases has on the human body can all too often be seen when people have died in fire situations. Experiments conducted on individuals have demonstrated very clearly the danger that this combination presents. If a group of individuals were to be blindfolded in their workplace and asked to find their way to open air, the majority would be able to do this with little problem, provided they had worked in that area for a reasonable period of time. However, if a very small amount of smoke containing a mixture of the chemicals mentioned above is introduced into their bodies, the situation will be very different. These chemicals target the central nervous system and produce a disorientating effect on the body. People breathing such material will, in a very short time, be confused, resulting in panic. This has the effect of increasing the body's breathing rate, which, in turn, results in more smoke being taken into the body.

Fire victims seldom die of actual burns. The majority die from the effects that breathing such gases has on the body. Confusion, disorientation and panic cause a person to struggle to find their way out of the room they were working in or sleeping in. Unfortunately, many victims are found very close to points of exit.

THE FIRE TRIANGLE

For a fire to start and to continue burning, it needs three things:

- A fuel – something to burn, such as a flammable solid, gas or vapour;
- oxygen – usually from the surrounding air, and
- an ignition source – in the form of a spark or naked flame.

Fuel

Fuel for a fire takes the form of:

- flammable solids, such as paper, wood, plastics and fabrics,
- flammable liquids, such as petrol,
- flammable gases, e.g. methane and propane.

However, the situation is a little more complex than simply saying that a fuel is a solid, liquid or gas. In fact, solid materials, or liquids, do not actually burn! When wood or paper is involved in a fire, they first decompose, producing carbon products. It is these carbon products that then combine with oxygen and ignite if a suitable source of ignition is applied. For example, holding a match up to a piece of paper does not immediately set the paper alight. The first thing that can be observed is some charring or darkening of the paper, which then produces a vapour product. It is the vapour that actually ignites if the heat source remains in place. This process of decomposition is called *pyrolysis*.

Pyrolysis is an important topic as far as combustion is concerned. Several serious fires have started in historical premises when timber joists, hidden underneath flooring, have been subjected to heat from exposed heating pipes. The heat from the pipes caused the timber to dry out and, over a period of time – in some cases, many years – the process of pyrolysis developed. The timber started to decompose and eventually gave off enough flammable vapour to ignite spontaneously, starting a fire in turn.

As in the case of solids, liquids also do not burn. Some liquids will, however, give off a flammable vapour, which does burn if a sufficient heat source is available. Whether or not a liquid produces flammable vapour depends on several factors.

Flashpoint

Flashpoint can be defined as the lowest temperature at which a substance will flash momentarily if a flame or heat source is applied. For example, the flashpoint of methylated spirit is approximately 10 °C. If a beaker contained methylated spirit in a room where the temperature was, for example, 5 °C and a naked flame was held slightly above the surface of the liquid, nothing would happen. The vapour from the liquid would not ignite.

However, if the beaker of methylated spirit were taken into a room with an ambient temperature above 10 °C and a flame was applied, there would be a momentary flash of heat and light from the surface of the liquid. Although this might be enough to spread onto another combustible material, the actual flame within the beaker of methylated spirit would go out after it had flashed momentarily. The liquid in this case would have reached its flashpoint.

The term *flashpoint* is an important one as far as legislation is concerned. Under the 'Highly Flammable Liquids and Liquefied Petroleum Gases Regulations 1972', a substance with a flashpoint below 32 °C is designated as 'highly flammable'. If such substances are used in the workplace, employers must comply with the requirements of these regulations, which are dealt with in a later chapter.

Clearly, liquids with flashpoints below ambient temperature pose a hazard, as they will always produce a flammable vapour. From a fire risk assessment point of view, these substances require proper storage arrangements, suitable fire-fighting equipment to be located nearby, properly trained employees and adequate signs and markings for containers and storage areas. Some common substances and their flashpoints are listed in Table 8.1.

Flammable Range

Another important consideration as to whether a substance will burn readily is the flammable range of the material.

Many explosions have occurred in domestic, commercial and industrial premises as a result of mixtures of flammable gas and oxygen in the presence of heat sources. In one example, a householder returned home from work and smelled gas. She went into the kitchen and immediately turned on the light. As the gas appeared to be coming from the cooker, she decided to call the fire

Table 8.1

Substance	Flashpoint (°C)
Diesel oil	+40
Acetone	−17
Petrol	−40
Methylated spirit	10
Toluene	4

brigade. Meanwhile, she opened the kitchen window to ventilate the room. As she left the room, she turned off the light, which resulted in a massive explosion, which destroyed most of the house.

The light switch caused a spark large enough to ignite the gas–air mixture. However, why did the gas not ignite when the occupier turned on the light? Instead, the gas mixture ignited as she turned off the light.

Clearly, on entering the room, the gas–air mixture was not actually flammable. However, after opening the window and allowing enough air to dilute the gas, the mixture became flammable and ignited as she turned off the light. This concept of flammability and, more importantly, the variation in flammability, is extremely important in the workplace and may be an important factor in a fire risk assessment.

Consider the combustion of methane gas in air. The reaction can be expressed by the following equation:

$$CH_4 + O_2 = CO_2 + H_2O$$

If this reaction were to be carried out in controlled conditions within a laboratory, precise amounts of each gas would have to be mixed together and the mixture subjected to a heat source for the above reaction to occur. In order to examine more clearly the exact amounts of each gas needed for complete combustion, it is necessary to balance the equation. A balanced equation for the above reaction would be as follows:

$$CH_4 + 2O_2 = CO_2 + 2H_2O$$

This means that one volume of CH_4 (methane) has to be reacted with two volumes of oxygen to produce one volume of carbon dioxide and two volumes of water (in the form of steam).

Now, let us consider the situation outside the laboratory, because in the workplace it would be impossible to measure out precise amounts of each gas. In the majority of cases, air contains about 20% oxygen, so complete combustion would occur if one volume of methane reacted with ten volumes of air:

$$CH_4(1\text{ volume}) + 2O_2(10\text{ volumes}) = CO_2 + 2H_2O$$

This exact mixture of gas and air produces complete combustion and is often referred to as the stoichiometric value. However, as previously stated, in the workplace this exact mixture will not usually be found. What is important in the workplace is the range either side of the stoichiometric value at which the gas will ignite.

Two terms are used to provide an indication of the flammable range of substances. These are the lower flammable limit (LEL) and the upper flammable limit (UEL). In the case of methane, the LEL is 5% and the UEL is 15%. In practice, this means that a room filled with methane in a concentration of between 5 and 15% will result in a flammable mixture with air and would result in an explosion if a flame or other ignition source were applied.

Using the previously discussed case of the kitchen explosion, when the occupier entered the kitchen and turned on the light, the gas in the room was probably at a concentration greater than 15% and therefore was too rich to ignite. However, by allowing air into the room, this would have had the effect of diluting the methane. The percentage of gas in the room would have dropped probably into its flammable range, i.e. between 5 and 15%. Thus, when the occupier turned off the light switch, a spark ignited the flammable mixture.

The above example illustrates how important it is to properly assess risks from flammable gases and to put in place proper control measures to reduce the risk. As part of the risk assessment, if employees are using or storing flammable gases, data on the flammability and on the flashpoint should be sought from suppliers. As a result of the assessment, training and procedures should be implemented to control the risk from these activities. Some examples of the flammable range of substances are given in Table 8.2.

Ignition Sources

The energy required to start the combustion process comes in many different forms. Some possible sources are as follows:

Naked Flames

Candles, matches, blowtorches are all examples of sources of heat that might start a fire. This method is usually referred to as direct burning, from one material to another through flame contact.

Another method of providing sufficient ignition energy through direct flame is through radiated heat. A common material ignited in this way is timber. If a fire is burning nearby, infrared rays from the fire may fall onto unprotected timber. The timber absorbs the rays and in turn releases energy through the process of decomposition and pyrolysis. The fumes and gases produced by

Table 8.2

Fuel	Lower explosive limit (%)	Upper explosive limit (%)
Hydrogen	4	74
Carbon monoxide	12	74
Methane	5	15
Propane	2.5	9.5
Butane	2	8
Acetylene	2.5	80
Petrol	1	7.5

the decomposition process may continue to heat, and, eventually, there will be sufficient energy to cause ignition.

Contact with a Hot Surface

A fuel, in the form of a liquid or a solid material, may fall directly onto a hot surface and ignite. A common example of this is where the insulation around hot steam pipes has broken down, exposing the pipe surfaces. Because of poor housekeeping, leaks of oil or other flammable liquids may fall onto the surface of these hot pipes, leading to ignition.

Other hot surfaces include those that are due to:

- friction between hot or seized bearings in machinery,
- general overheating of equipment caused by the failure of thermostats,
- overheating of electrical equipment caused by overload or inadequate circuit protection.

Arcing

Electrical equipment may produce an arc or spark from:

- the breakdown of insulation, allowing an arc between the live cable and an adjacent earth,
- arcing at contacts, e.g. brushes of motors or contacts of thermostats,
- the operation of switches.

All the above may produce sufficient energy to start the combustion process unless adequate precautions are taken.

Electrostatic Discharge

Most materials, when moving against other surfaces, create static energy. The energy produced may be enough to initiate a combustion reaction. Examples of static discharge include the following:

- The movement of solids, e.g. powders, grain and other dusts, through ducting or chutes will generate charges. These charges, unless discharged away from the area, will be sufficient to ignite flammable material nearby.
- The movement of liquids through pipes will also generate charges of sufficient energy to initiate a combustion reaction. Fires that are generally caused by this method have occurred during tank-filling operations. If petrol is decanted from a metal container into another metal tank, there might be enough build-up of static energy on the two metal surfaces to ignite the petrol vapours should a spark occur. An appropriate control measure in this case is to use a plastic funnel to provide an insulation barrier between the two metal surfaces.
- People can also generate static energy. Static energy can build up on clothing and also on flooring when people walk over it. Special clothing may be required when employees are operating near flammable atmospheres.

Spontaneous Combustion

There are many processes in which the risk of spontaneous combustion is present. This type of combustion usually involves a self-heating process, until eventually ignition occurs. Typical examples of situations in which the risk of spontaneous combustion is present are as follows:

Compost heaps Organic matter usually undergoes a self-heating reaction, especially if there is a high level of moisture. The moisture helps to support microbial activity, which then results in heat being produced. If the heat is trapped within the compost heap, the heat builds up to a point whereby the organic material can ignite (if it reaches its self-ignition temperature).

Wood sanding byproducts Gloss paint contains drying oils, which harden when exposed to oxygen. However, some of the drying oils remain unreacted and become trapped within the underlayer of dried and hardened paint.

Should sanding operations take place, the unreacted drying oils are then released into the atmosphere along with wood sandings. If the wood sandings are swept into a heap, which helps to insulate the oils, a self-heating reaction may occur. The heat build-up is trapped within the insulated sandings, and, eventually, the organic material may ignite.

Damp hay In damp hay, biological activity produces heat, which is once again trapped within the stack. As the temperature increases, biological activity also increases, resulting in more heat being produced. Provided the stacks of hay are kept small enough, there will be enough air circulation to reduce the heat build-up. However, where stacks are large enough, air movement may be limited, causing a significant level of heat to build up within the stack. Spontaneous combustion is then possible.

Other high-risk situations as far as spontaneous combustion is concerned include the following:

- Large coal heaps.
- Paint or oil-soaked rags.
- Laundry.
- Fertilizers, particularly those based on ammonium nitrate.
- Bulk paper storage, especially recycled paper bales.

Control measures to reduce the risk of spontaneous combustion are as follows:

- Carrying out risk assessments to determine stack size and possible heat build-up.
- Controlling moisture content.
- Checking temperatures within the stacks.
- Reducing the size of the stacks below their critical size.
- Control of external ignition sources, e.g. sparks from vehicles, hot work, fireworks, static electricity and so forth.

EXTINCTION OF FIRE

Once a fire has started, extinction can be achieved by a number of different methods, mainly through removal of fuel, oxygen or heat.

Removal of Fuel

The simplest method of stopping a fire is the removal of its fuel. Without something to burn, a fire cannot possibly exist. Although this might seem simple, how can it be achieved in practice?

Clearly, this method of extinguishing a fire is better dealt with as a preventative measure rather than as a cure once a fire has started. A simple example of this is to create firebreaks in crop fields and in forests, which, if a fire should start, will eventually burn out after the available fuel has been used.

Firebreaks are not just confined to outdoor situations; it is equally important to consider these where the storage of goods indoors is concerned. Warehouses, partially covered yards and other bulk areas should be designed with firebreaks in mind.

Where firebreaks may be impossible, another control measure to limit the fuel available for a fire is to create fire barriers. One example that is used extensively in the petrochemical industry is to install drencher systems to protect one area of a site from another area. Should a fire start in a part of the plant, a drencher system automatically provides a curtain of water around other areas where flammable materials are used or stored.

Removal of Oxygen

The simplest example of this method of fire extinction is the common chip pan fire. By carefully placing a damp cloth over the top of the pan, oxygen is prevented from reacting with the oil vapour. This has the effect of extinguishing the fire. Of course, as an added precaution, the heat source should also be removed.

The removal of oxygen is the basic method of extinguishing all flammable liquid fires. The process of combustion when a flammable liquid burns is as follows:

- The flammable liquid heats up, forming a flammable vapour. This sometimes occurs after heating and sometimes naturally, depending on the flashpoint of the liquid.
- The flammable vapour mixes with oxygen in the air to form a flammable mixture, depending on the flammable range of the mixture.
- Ignition from a heat source initiates combustion.
- The flame emits radiant energy, which in turn heats the liquid and produces more vapour.

To extinguish a flammable liquid fire, the above chain must be interrupted by some method, usually by cutting off the supply of oxygen to the vapour. This is done in two main ways: by physical means and by displacement.

Physical Means of Removing Oxygen

Just as in the example of the chip pan fire, a physical barrier can be placed on top of burning liquids to smother the material. Another example of placing a physical barrier is through the use of a foam extinguisher. To be effective, foam must have a number of properties:

- It must be able to flow over the surface of the burning material and thus cut off the supply of oxygen.
- The foam must be resistant to the vapour produced by the flammable liquid. Standard foam extinguishers are not capable of producing foam that is resistant to some vapours.
- The foam must be resistant to contamination by the liquid. If this occurs, the foam could break down, resulting in possible reignition of the vapour.
- The foam must be resistant to heat, so radiant heat from the flame will not cause the foam layer to break down.

There are two types of general foam used for extinguishing purposes and it is important that, as a result of fire risk assessment, the best type for the circumstances is determined.

Protein-based foam Protein-based foam has been used extensively for a number of years and provides a good extinguishing medium. However, one of the disadvantages is that it provides little in the way of cooling the burning liquid. This lack of cooling, in some situations, may result in the foam only being effective for a short period before reignition occurs.

Aqueous film-forming foam (AFFF) This foam has a special agent that allows a water film or a foam film to float across the surface of the burning liquid in front of the actual foam layer. This film provides a seal, which helps reduce vaporization of the liquid and allows for better extinction times.

Displacement

Another method of excluding air from a fire is by displacement. This method of fire extinction involves replacing the oxygen content of an area by replacing it with a gas that will not support combustion. Such gases include carbon dioxide, nitrogen, steam and other inert gases.

In general, flame propagation will usually stop and, hence, a fire will go out if the oxygen content in the area is reduced to below 12%. One of the problems with using displacement as a fire-extinguishing method is that often quite large quantities of inert gas must be present before combustion ceases. In the case of carbon dioxide, an area would require about 60% of its volume to be filled with this gas before the atmosphere ceases to support combustion.

Removal of Heat

The most common method of extinguishing most fires is by heat removal or cooling. Water is the most effective cooling agent as:

- it requires large quantities of heat to raise its temperature,
- it requires even larger quantities of heat to convert it to steam,
- it will usually absorb large quantities of heat if it reacts with the fuel.

Water may be applied onto a fire in a number of different ways:

Water Applied as a Jet

For large fires, the most effective method of extinguishing them is to apply large amounts of water in jet form. Most water extinguishers will produce a small jet, which can be aimed at the heart of a fire and will exert an immediate cooling effect.

Water Applied as a Spray

A much more effective method of extinguishing fires, particularly small fires, is to apply water in spray form. This is effective because the droplets produced by the spray have higher surface areas, which in turn absorbs more heat from the fire. In addition, water in spray form will limit the amount of water damage to the rest of the premises.

Other Methods of Extinguishing Fires

Water, carbon dioxide and foam all operate in their own way to reduce the heat and oxygen in a fire. However, there is another method of fire extinction that does not really involve the removal of heat, fuel or oxygen.

The use of dry powders as a fire-extinguishing medium does not cool a fire (because it is in powder form), it certainly cannot remove the fuel and it does nothing in terms of the oxygen because it cannot form a blanket over the fire. Yet, dry powders are an extremely effective method of extinguishing a fire.

Dry powders are classed as negative catalysts as they absorb the energy from the molecules that are involved in flame propagation. If the energy of these molecules is absorbed, they cannot react further; therefore, combustion ceases.

Another agent that is even more effective than dry powder is vaporizing liquid (halons). These do not remove heat, fuel or oxygen but react in a very similar way as dry powders; they remove the energy of the flame propagation molecules.

A Summary of the Different Methods of Fire Extinction

Type of extinguishing medium	Method of operation	Use
Water	Removal of heat by cooling the fire.	General fires – but not fires involving live electrical equipment.
Foam	Removal of oxygen by placing a blanket over the burning fuel.	Flammable liquids and general fires – but not fires involving live electrical equipment.
Carbon dioxide	Displacement of oxygen.	Fires involving live electrical equipment (CO_2 in a non-conductor of electricity) and small flammable liquid fires.
Dry powder	Interference with the chemical reaction of flame propagation.	Fires involving live electrical equipment and small flammable liquid fires.
Vaporizing liquids	Interference with the chemical reaction of flame propagation.	Fires involving live electrical equipment.

FIRE SPREAD

Another important point to consider whilst carrying out fire risk assessments is the likelihood that fire will spread throughout the premises or to adjoining premises or plant.

Fires can spread by a combination of factors:

- Heat transfer e.g. radiation, conduction and convection.
- Direct burning through flame contact.
- The release of flammable gases.

Heat Transfer – Radiation

Heat transmission from invisible rays, which travel through air, occurs from hot surfaces and from fires already burning. A common example of this method of fire spread is clothes hanging near an electric heater. As the clothes dry out, they absorb the energy in the form of rays from the electric heater until sufficient energy has been absorbed and combustion is initiated.

Fire can spread through the process of radiation to nearby buildings, especially old buildings that are built closely. It is possible that a fire starting in one building could generate enough energy within the invisible rays to start a fire in an adjoining building. Often, insurance companies will insist that water curtains be installed in old buildings to cool them in the event of a fire starting elsewhere.

Heat Transfer – Conduction

Conduction involves heat passing through a solid from one area to the next. A poker held in an open fire soon gets very hot at the tip. However, as the heat travels along the poker, by the process of conduction, the other end of the poker also gets very hot.

Fires in buildings may spread by conduction through a steel door or along a steel beam. Part of the building regulations now require beams that penetrate through fire walls to be cased in fire-resisting material.

Heat Transfer – Convection

Hot gases and smoke rise from a fire. As they cool down, they become denser and so fall back to ground level. This circulation system is the basis of central heating radiators used for heating purposes.

A significant fire risk is also possible through the process of heat convection. If a fire starts in a building, for example in a lift shaft, the hot gases will rise from the fire and move to high levels. These hot gases contain fuel, usually in the form of unburned carbon coming from the fire, and heat. All the gases lack is a fresh supply of oxygen. If a person were to open a door in the lift shaft and allow a sudden flow of air to react with the hot gases, it is possible that a violent explosion could occur, thus spreading the fire from its original location to upper areas of the premises. Stairwells, ducting, refuge shafts or any other open and vertical area is particularly at risk from this type of fire spread.

Direct Burning

The most common method of fire spread in buildings and in the workplace is by direct burning. A fire will easily and quickly spread to other materials if they are present. Good housekeeping can prevent a fire from spreading from its simple beginnings to a more serious incident.

The Release of Flammable Gases

One of the most underestimated hazards in fires is the fact that a burning fire produces large amounts of hot flammable gases. Many people have been killed or injured as a result of this phenomenon. Two terms are particularly important and their consequences should be assessed during the process of fire risk assessment. These terms are:

- flashover, and
- back-draught.

A typical indoor fire progresses through three main stages, and it is possible to model this development.

Stage One – Initiation

As previously mentioned, for a fire to start it requires a source of heat to start the combustion process.

Stage Two – Rapid Fire Growth

During this stage of the fire, the output in terms of heat grows quickly, as there is plenty of fresh oxygen and fuel for the fire to burn. At a certain point in the fire growth, most of the contents of the room that is on fire will reach their self-ignition temperature and will ignite spontaneously. At this point, the room will suddenly become engulfed in flames, with everything in the room igniting at once. This point is called 'flashover'.

In domestic premises, it is often the noise created during this stage that people hear first. Breaking of glass, small mini-explosions as electrical equipment burn and cracking noises are usually experienced during this stage of fire development.

Stage Three – Decelerating growth

After the point where all the contents of the room ignite, the fire quickly gets starved of its supply of oxygen and the heat output starts to tail off. During this stage of the fire, large amounts of unburned gas, in the form of carbon monoxide and solid particles of carbon, are produced and rise upwards. These gases are superheated and are potentially very explosive. However, the gases will not explode because most of the available oxygen in the compartment will be depleted at this point. This is the most hazardous of the stages of fire development because if a person enters the room or a window is broken, the fire will draw the fresh air in and mix with the gases to produce a highly flammable mixture, which may explode violently. This type of explosion is called a *back-draught*.

While carrying out fire risk assessments, it is important to identify the possible stages of fire development and the possible routes of drawing in fresh air, either accidentally or intentionally. Accidental ventilation may occur when someone smashes a window to try and clear the smoke and fumes or when a door is opened to re-enter the building.

If these factors are likely to occur, suitable training should be given to employees and others who may be at risk.

SPONTANEOUS COMBUSTION

Certain substances react with air or moisture, liberating heat in the process. If the heat is not dissipated quickly enough into the surroundings, the temperature of the substance will increase. If this increase in temperature reaches the substance's self-ignition temperature, the material may ignite, even in the absence of a flame or ignition source.

Some examples of the most commonly used materials that display spontaneously combustible characteristics are given below.

Metals

Finely divided metals have a large surface area, which is available for reaction with air or water. Some metals cause a decomposition reaction with water, with the result that heat is produced. Other metals may react slowly with air so that an oxide coating is produced, which then becomes less reactive.

Pyrophoric Gases

These substances are quite rare; however, they are used extensively in the semiconductor industry. Examples include the following:

- Diborane
- Phosphine
- Silane

Water-Reactive Substances

Some substances react with air moisture to produce flammable gases such as hydrogen, acetylene and ammonia. If the heat generated is sufficient, these gases may ignite spontaneously. Examples include the following:

- Reactive metals
- Borohydrides
- Hydrides
- Carbides
- Phosphides

Phosphorus

Phosphorus usually comes in two forms: red phosphorus, which reacts slowing with air and tends to be reasonably stable, and white phosphorus, which reacts quickly with air and is therefore stored under water.

CONTROL MEASURES FOR SPONTANEOUSLY COMBUSTIBLE SUBSTANCES

The following general control measures should be used when spontaneously combustible substances are stored or used in a workplace.

Information and Instruction

Specific information regarding the storage and use of these substances can be obtained from the manufacturer's data sheets. This information should be passed on to all employees and others who work with such substances.

Storage

The following general storage arrangements should be considered when storing spontaneously combustible substances:

- Spontaneously combustible substances should be stored separately from other dangerous materials.
- Substances should be stored in areas that are well away from buildings or vehicle movement.
- If these substances are stored in the open, weather protection may be necessary to prevent corrosion of containers.
- Depending on the type of material that is being stored, e.g. pyrophoric substances, the integrity of the storage vessel should be checked regularly to ensure that no damage has occurred and that the container is intact.
- Reduce the amount of substance that is stored to an absolute minimum.
- Do not pack containers of substances too closely together.

Purging and Cleaning of Storage Vessels

Vessels and containers used to store spontaneously combustible substances should be thoroughly cleaned and dried before use. Purging may also be necessary before introducing a pyrophoric substance.

Reaction with Water

Some substances react with water, producing flammable gases in the process. When these substances are stored or used, care should be taken to ensure that water cannot ingress into containers or flow lines. Procedures to ensure that water moisture is removed before these substances are allowed to enter is important.

Shelf-life

Some substances can start to decompose if left in storage for long periods. Care should be taken to rotate stock and to ensure that substances beyond their shelf-life are disposed off correctly.

Fire Risk Assessment

When storing or using spontaneously combustible substances, a fire risk assessment should be carried out before the operation begins or before the substances are taken into storage. The following points should be considered:

- Type of substance.
- Amount required for a given task or for a given period of time.
- Storage arrangements currently in place and their suitability.
- Training of personnel who are likely to handle these substances.
- Reactivity of the substances.
- Shelf-life and decomposition characteristics.
- Self-ignition temperature of the substances.
- Reactivity with air or moisture.
- Reactivity with other substances.
- Purging and cleaning of containers.
- Safe systems of work or permit-to-work systems that may be necessary to control cleaning operations.

9 Explosions

INTRODUCTION

An explosion is a sudden release of energy, causing a pressure blast wave. It is often this blast wave that causes damage to buildings and other property. Also, it is the pressure wave that often causes physical harm to people as a result of flying debris.

Explosions can occur during a number of different activities. Some examples of these are as follows:

- Escapes of gas or flammable liquids in the open may provide a large gas or vapour cloud that ignites if a flame, spark or other source of ignition is applied.
- Petrol is one of the most common examples of a substance with a high risk of explosion. Petrol vaporizes readily at normal ambient temperatures, forming a heavy gas. Given the right conditions, i.e. a suitable ignition source, the flammable gas can ignite and produce significant pressure wave damage.
- During discharge or filling operations in which flammable materials are involved, a significant risk of explosion is present. Static electricity is a common source of ignition where these operations are taking place.
- During hot work, e.g. welding, cutting or grinding near tanks that contain flammable liquids or gases.
- Explosions in small pleasure vehicles, e.g. boats, caravans and camping equipment, have often resulted from poor handling, storage and use of propane or butane gas.
- Explosive chemicals, which have been mixed together, either deliberately or accidentally.

There are a number of different definitions of explosion, depending on the nature of the material involved. The most common types of explosion mechanism involve one, or more, of the following:

- Deflagrations
- Detonation
- Confined vapour cloud explosions
- Unconfined vapour cloud explosions
- Boiling liquid expanding vapour cloud explosions
- Chemical explosions
- Dust explosions.

DEFLAGRATIONS

A deflagration-type explosion occurs when a flammable gas or vapour becomes mixed with air to form a flammable air–gas mixture and this resultant mixture is exposed to an ignition source. A deflagration may involve a number of different materials, e.g. propane, butane and methane.

A deflagration produces a massive shock wave; usually, it is the shock wave that results in property damage. It is possible for the shock wave to propel very heavy items hundreds of metres through the air, thus causing massive structural damage and injury to people.

If there is a gas–air mixture within a building or premises, and an ignition source is present, the deflagrating reaction propagates through the mixture, increasing the temperature as it travels. The increase in temperature is accompanied by an increase in pressure, which builds up rapidly. Although the increase in temperature is a problem, it is the pressure rise that is most concerning. Deflagration shock waves from the pressure can travel through the air at subsonic velocities and can cause considerable structural damage. An internal explosion of this type is often much more serious than an external one, especially where structural damage is concerned.

DETONATION

Detonation can be even more destructive than deflagration-type explosions as the resulting shock wave travels much faster, often at supersonic velocities.

A detonation is an extremely rapid, self-propagating decomposition of an explosive material. In the majority of workplace situations, it is unlikely that detonation explosions will occur. This is because the type of material that causes such explosions is usually designed to explode and not to perform general workplace tasks.

CONFINED VAPOUR CLOUD EXPLOSIONS

If a flammable gas or vapour in the right concentration with air is present within a container or within the confines of a building, ignition by a heat source can cause pressure to build up rapidly. This deflagrating reaction causes the walls of the vessel or the building to collapse under pressure. This type of explosion is responsible for the majority of workplace accidents that involve explosive materials.

Confined vapour cloud explosions can cause considerable structural damage, with or without fire. The effects of such an explosion are usually localized to the area immediately around the building or the container; however, this is often enough to cause debris and blast damage sufficient to inflict considerable injury to people nearby. Many fatalities have resulted from explosions within buildings or containers, particularly during filling or decanting operations.

UNCONFINED VAPOUR CLOUD EXPLOSIONS

This type of explosion occurs when a large quantity of flammable gas or vapour is released into the atmosphere, without being confined to a building or container, and is subjected to an ignition source. Unlike a confined vapour cloud explosion, the unconfined cloud explosion produces considerable blast damage in a wide area around the initial explosion.

Many gases and vapours are heavier than air and can travel considerable distances from the source of origin to the point of ignition. In addition, gas clouds can sink into basements, drains or other vents and ducts and can present a significant risk of explosion up to some distance from the point of origin.

Some common examples of situations in which a gas might escape include the following:

- During filling and decanting operations.
- Leaks from pipework.
- Gas used for portable heating and cooking equipment, which may leak from the flexible hoses.
- Poor maintenance around flexible couplings and valves, causing leaks.
- Inadequate instructions for changing gas cylinders.

BOILING LIQUID EXPANDING VAPOUR EXPLOSIONS (BLEVEs)

One of the most significant underestimated hazards in the workplace is from gas cylinders. Often, these are used for heating appliances, cooking, welding and other hot work. In addition, they can be used as a fuel source for fork-lift trucks and even cars and vans.

If used properly, gas cylinders present little risk; however, if used and stored incorrectly, they can present a major hazard.

A BLEVE involves a sudden release of vapour, usually from portable gas cylinders or bulk tanks containing liquefied petroleum gas (LPG). A BLEVE occurs when the vessel containing such a substance is heated so that the metal container loses strength and ruptures. The rupture causes a release of vapour, which then ignites if a source of ignition is available.

The chain reaction of a BLEVE often consists of the following sequence of events:

1. A fire starts near a cylinder that contains LPG.
2. The flames impinge on the metal walls of the cylinder, causing it to heat.
3. The hot cylinder transmits heat to the liquid inside, causing the liquid to heat up.
4. If the flames are not extinguished, the liquid undergoes a reaction whereby the liquid changes to a gas, causing massive overpressure inside the cylinder.
5. Emergency vents in the cylinder blow off, releasing the pressure. If the flames continue to impinge on the cylinder, the gas inside continues to heat, raising the pressure even more. If this happens, the blow-off valves cannot release enough pressure.

Table 9.1

Gas	Critical temperature
Propane	97 °C
Butane	153 °C

6. The cylinder eventually ruptures. It can take between 10 and 20 minutes for a cylinder to finally explode.
7. The release of vapour from the cylinder then ignites, owing to the flames, and this causes a massive deflagration-type explosion.
8. The shock wave produced by this type of deflagration is enough to cause debris to fly off in all directions. It has been estimated that a 15 m diameter sphere of LPG could result in blast damage up to 5000 m from the container. Burns have been inflicted upon people up to 1000 m from a gas cylinder after it BLEVEs.

This type of explosion can occur from any liquefied gas container. The crucial part of the sequence of events leading up to explosion is the critical temperature. This temperature is defined as 'the temperature above which the substance cannot remain as a liquid no matter how high the pressure is'. Normally, these gases are stored in liquid form; however, when heat starts to raise the temperature above the liquid's critical temperature, the liquid changes to a gas.

Some common gases and their critical temperatures are provided in Table 9.1.

It might be assumed that the aforementioned temperatures are very high and would never be reached under normal conditions. However, if an LPG cylinder were to be left in a room that was involved in a fire, it is likely that the temperature within the room would reach about 350 °C after the first 4 minutes of burning. A gas cylinder in such conditions would be most likely to BLEVE.

CHEMICAL EXPLOSIONS

There are a number of chemicals designed as explosive materials; however, there are also a large range of chemicals that have explosive properties as a by-product of their main design.

The following are some of the most common types of chemical in general use that have explosive properties.

Acetylene Gas

This gas is used mainly for welding and cutting operations. The gas, when used properly presents little risk, but when it is stored or used incorrectly, the risk of explosion is significant. Acetylene may decompose, releasing energy rapidly if subjected to excessive heat, e.g. if a flame impinges on the cylinder or if it is stored near hot surfaces. Acetylene gas may even decompose if the cylinder has been handled roughly, e.g. if it has been knocked or dropped on a hard surface. Once decomposition starts, it may take up to 24 hours before the gas generates enough heat to explode.

Ammonium Nitrate

This substance, which is generally used as a fertilizer, has explosive properties when mixed with other substances. Large-scale explosions have occurred on farms where quantities of ammonium nitrate have been stored in barns. Often, if the substance is mixed with oil, a self-heating reaction occurs, causing a rapid build-up of heat and decomposition of the nitrate. This releases nitrogen dioxide gas and in the process may produce a massive blast wave.

Sodium Chlorate

This material is used as a weedkiller and has been known to explode when involved in a fire situation. The substance does not behave in the same manner as nitrates but certainly displays extremely high explosive properties.

Organic Peroxides

These are often used as catalysts during plastic manufacture. The substance explodes violently if it is involved in a fire.

Energetic Substances

An energetic substance is one that releases energy under certain circumstances. These substances are capable of undergoing rapid exothermic decomposition, which may result in a fire or an explosion. Energetic substances may be:

- unstable at or below normal room temperature,
- stable at normal room temperature but react with air or moisture,
- stable at room temperature but decompose on heating, possibly after a prolonged period of time.

While carrying out fire risk assessments, all chemicals stored or used in the workplace should be assessed for their fire or explosive properties. During this process, chemicals should also be assessed as to their energetic characteristics, which will certainly result in separate storage arrangements and use. Decomposition of these substances may occur during various operations and therefore it is not the mixing of chemicals alone that must be assessed. Fire risk assessment should be carried out under the following circumstances:

- Mixing (deliberate or accidental) of energetic substances.
- Mixing (deliberate or accidental) of energetic substances and non-energetic substances.

- Storage of raw materials that are classed as energetic substances.
- Storage of finished products that contain energetic substances.
- During distillation operations.
- Manufacturing and processing.
- Transporting.

The following set of questions may need to be asked to ascertain whether a chemical can be classed as having 'energetic' properties. If the answer to any of these questions is 'yes', the chemical probably can be classed as a high-risk chemical:

- Does the label on the chemical indicate that it may be 'explosive'?
- Is the substance classified as one belonging to 'Class 4'?
- Is the substance classified as one belonging to 'Class 5'?
- Does the chemical structure (probably provided on the manufacturer's or supplier's data sheet) indicate that it has possible energetic properties?
- Does the data sheet for the chemical indicate reactive properties?
- Does the data sheet for the chemical indicate instability?

LABELLING OF SUBSTANCES

There are two important classifications that determine the type of label that a substance should have. These simple labels are a good source of information as part of a fire risk assessment. The two classification schemes are as follows:

Carriage of Dangerous Goods by Road and Rail (CPL) Regulations 1994

These regulations apply to substances that are transported by road or by rail. Substances are classified into one of the nine groups described in Table 9.2 according to their principal hazards:

Chemicals (Hazard Information and Packaging for Supply) Regulations 1994

These regulations impose a duty on the suppliers of substances, which must be classified according to their hazards and risks and must be labelled accordingly.

There are three main categories, and a further 15 subcategories, of risk from substances that fall within these regulations.

- Physico-chemical
  - Explosive
  - Oxidizing
  - Extremely flammable
  - Highly flammable
  - Flammable

Table 9.2

Classification	Description	Examples
Class 1	Explosive substance	A substance that is capable, by chemical reaction, of producing gas at such a temperature, pressure and such a speed that it could cause damage to the surroundings.
Class 2	Gas	A substance that is completely gaseous at 20 °C.
Class 3	Flammable liquid	A liquid with a flashpoint below 51 °C.
Class 4.1	Flammable solid	A solid that is steadily combustible or may contribute towards a fire.
Class 4.2	Spontaneously combustible substance	A substance that is liable to spontaneous combustion under certain conditions.
Class 4.3	Substances, which in contact with water, emit flammable gases.	A substance, which in contact with water, is liable to become spontaneously combustible or to give off a flammable gas.
Class 5.1	Oxidizing substance	A substance other than organic peroxide, which may yield oxygen and therefore contribute to combustion.
Class 5.2	Organic peroxides	A substance that may undergo exothermic self-accelerating decomposition.
Class 6.1	Toxic substances	A substance that is liable to cause death, serious injury or harm to human health if swallowed, inhaled or absorbed through the skin.
Class 6.2	Infectious substance	A substance that contains viable microorganisms that are known or believed to cause disease in animals or humans.
Class 7	Radioactive material	A substance that meets the criteria of the Radioactive Material (Road Transport) Act 1991.
Class 8	Corrosive substance	A substance that by chemical action will cause severe damage when in contact with living tissue.
Class 9	Other dangerous substances	A substance that may cause a risk to health or safety, but is not already covered by the above categories.

- Health
  - Very toxic
  - Toxic
  - Harmful
  - Corrosive
  - Irritant
  - Sensitizing
  - Carcinogen
  - Mutagenic
  - Toxic for reproduction
- Environmental
  - Dangerous for the environment

Should any of the above substances be used in the workplace, a general risk assessment must be carried out. From a fire assessment point of view, those substances marked with the following need to be assessed:

- Explosive
- Oxidizing
- Extremely flammable
- Highly flammable
- Flammable
- Dangerous for the environment (fire-fighting waters may cause pollution to the watercourse or nearby rivers).

CHEMICAL STRUCTURE

The manufacturer's or supplier's data sheets may contain information relating to the chemical composition of a substance. This information is extremely valuable when assessing the risk from substances. Often, this is an indication of the degree of energetic characteristics that the substance may exhibit.

Reactive or unstable Properties

Many energetic substances react with other substances, sometimes in a violent manner. For example, one of the largest explosions and resultant fires occurred at the Allied Colloids chemical plant in Bradford in 1992. Two substances, an oxidizing agent called sodium persulphate and a reducing agent called azodiisobutyronitrile (AZDN), were incorrectly stored together in the same storage area. A sequence of events occurred, enabling the substances to become mixed, and a spark ignited the mixture. The resulting fire caused approximately £6 million of damage to the facility.

Some energetic substances can become unstable if left for long periods. Regular checking of stock levels and shelf-life information is therefore an essential part of the risk management process.

CONTROL OF RISKS

Risk control will largely depend on the types of energetic substances that are stored or used within the workplace. However, general guidance can be applied in almost all cases.

Storage

The following general precautions should be taken when storing and using energetic substances:

- Regular inspection of packaging for signs of damage, leakage or corrosion.
- Storage should be out of direct sunlight.
- Storage areas should be free from sudden temperature rises, e.g. unused heating systems that may become accidentally turned-on.
- Store frost-sensitive substances in different areas and ensure that an appropriate heating system is used to keep temperatures above freezing.
- Ensure that appropriate containers, which are recommended by the manufacturers and not substituted for 'makeshift' containers or storage vessels, are used.
- Do not store energetic substances with other chemicals.
- Do not smoke or use naked flames in storage areas.
- Use appropriate fire-fighting medium and ensure that adequate arrangements are made for sufficient quantities of fire extinguishers or bulk systems to protect the storage area.
- Make provisions for the collection of fire-fighting water, which may become contaminated as a result of fire-fighting operations.
- Plant and processes should be designed to ensure that the equipment does not produce sufficient heat to start the decomposition process.
- Ensure that sufficient data is available on the substances before they are used, including safe operating temperatures and upper and lower limits of decomposition.
- Ensure that the equipment that has become contaminated with energetic substances is frequently cleaned.
- Ensure that the equipment used to process energetic substances is frequently tested and maintained, e.g. heating elements should be operated within the normal ranges and too much or too little heat should not be applied accidentally.
- Ensure that suitable quality control procedures are in place to prevent inadequate quantities of substance being mixed and that correct flow rates are applied.

DUST EXPLOSIONS

Some dusts are capable of producing explosive reactions, which can be extremely destructive. There are four general factors that determine whether a dust is capable of exploding.

Characteristics of Dust

Some dusts exhibit explosive characteristics, i.e. they are capable of ignition if mixed with air. Not all dusts are flammable and, therefore, regardless of the air-dust mixture, they will not ignite. Whether dust is capable of exploding is dependent on the ignition energy required to initiate a combustion reaction and on its chemical components.

Dust is usually more difficult to ignite than flammable vapours. For example, it takes only about 0.02 mJ to ignite petrol vapour; however, a minimum of 5 mJ would be required to ignite the most explosive of dusts.

Size of Dust Particles

The dust particles must be of a suitable size. It is difficult to ignite large lumps of solid material, mainly because the surface area does not allow enough oxygen to become available. For example, it is difficult to ignite a sugar lump if a flame is provided. However, if the sugar lump is ground into very fine particles, the surface area of the total amount of solid material, now in dust form, increases substantially. It is now easier to ignite the substance.

Correct Flammable Range

The dust must be mixed with air within the flammable range of the substance. As with vapours and gases, the upper flammable limit (UEL) and lower flammable limit (LEL) are important factors when considering whether the dust will ignite.

Ignition Source

There must be a source of ignition, e.g. spark, flame, arc from electrical equipment or from static electricity.

The Sequence of Events of a Typical Dust Explosion

The following general sequence of events are normally involved in dust explosions, regardless of the material involved:

- A suitable amount of dust builds up on machinery or work surfaces.
- An ignition source is applied to the dust layer.
- The dust ignites and combustion rapidly spreads through the layer. This initial dust explosion is referred to as the primary explosion.
- The primary explosion causes the dust near the point of ignition to dislodge and causes some of the dust to fly into the air.
- The air disturbance disperses the surrounding dust, which then ignites as a result of the heat from the initial explosion. This is referred to as the secondary explosion.

Premises particularly at risk from dust explosions are those where the following processes are being undertaken:

- Where large quantities of aluminium powder is processed.
- Where the following materials are processed: sugar, flour or starch.
- Where grain is processed into animal or human foodstuffs.
- Where there is a large build-up of sawdust from sanding, cutting or grinding operations.

Control Measures

As a result of a fire risk assessment and, depending on the findings, the following general precautions may have to be taken.

Ignition Sources

Common sources of ignition include the following:

- Naked flames
- Friction sparks
- Sparks from electrical equipment
- Smoking materials
- Static electricity
- Sparks from hot work
- Sparks from poorly maintained machinery.

Naked flames should be controlled by safe systems of work or procedures and, where they are absolutely necessary in dust-laden areas, permit-to-work systems should be used to ensure that adequate control measures are put in place.

Moving parts of process equipment may produce sparks that can ignite flammable dusts. Modern equipment should be designed with this in mind; however, even with such equipment, maintenance is essential to ensure that frictional heat is not generated.

Often, a problem that is overlooked is frictional sparks produced from stones or other hard materials striking concrete or metal surfaces. Magnetic separators are often used to catch metallic materials before they can enter areas of grinding. Regular checks should be carried out to ensure that large trapped materials are removed from filters or other devices that are used to trap them. Sieves and other filtering equipment may have to be used to separate stones and other materials before they enter crushing and grinding equipment.

The design, installation and testing of electrical equipment used in dust-contaminated areas is similar to those used in flammable gas atmospheres. However, one of the problems with electrical equipment is when maintenance work has to be undertaken, especially when work is required at short notice.

Often, this last-minute type of work will result in items of portable electrical equipment being allowed into the working environment. For example, incidents have occurred where portable lighting, which is powered by mains supply, has been taken into a vessel containing combustible dust. With this type of equipment, the heat emitted from the item may cause the dust in the immediate area around the equipment to heat up, causing it to reach its explosive temperatures and to ignite spontaneously.

Care should be taken during emergency work, and safe systems of work and permits-to-work should be developed and implemented when such work is to be carried out. Contractors who may be called upon to do such work should be given adequate information about the systems and permits that exist to ensure that the task is carried out safely.

It may seem obvious that smoking materials should be banned from areas where combustible dusts are present. However, even if such procedures are in place, it is essential to maintain them. Supervisors and managers must ensure that the employees adhere to these procedures and should carry out regular inspections of the working environment.

Inerting

Where combustible dusts are used or produced as part of a process, the oxygen content of the area may need to be controlled, especially if ignition sources cannot be removed. By using an inert gas, the oxygen within the area where the dust is processed can be kept to a level below the LEL.

If inerting is used to control the atmosphere, constant monitoring of the oxygen level should be done and, should the oxygen level change and allow the dust to enter its flammable range, the operation of alarms and the shutdown of the plant should be automatic.

Control over Dust Cloud Formation

Probably the best form of control, as far as the potential for dust explosions is concerned, is to minimize or eliminate any dust formation in the first place. This can be achieved by:

- damping down the area concerned,
- using local exhaust ventilation systems,
- maintaining plant and equipment as leak-free as possible,
- providing planned preventative maintenance schedules for equipment and plant,
- carrying out checks of flanges, joints, flexible seals and so forth to ensure that they remain leak-free,
- ensuring good standards of housekeeping, e.g. regular cleaning for removing dust,
- cleaning higher, less obvious parts of buildings where dust has accumulated over a period of time.

Plant and Equipment

Process deviations should be controlled and it is often necessary to control such deviations by instrumentation, which is more reliable than simple visual monitoring. Process deviations may include the following:

- Maximum temperature of drying products before self-ignition temperatures are reached.
- Monitoring of the feed supply into the process equipment or plant.
- Monitoring of air pressure to ensure that air filtration equipment does not get blocked.
- Monitoring of local exhaust ventilation systems and flow and suction rates of airflow.
- Fire detection to monitor glowing material.

Human Factors

Even with the best control measures installed, dust explosions may still occur. In addition to the control measures mentioned previously, factors to control people also need to be put in place. These include the following:

- Information about the risks that dust explosions may present should be communicated to all employees working with processes in which dust is generated.
- Training should be given to employees who work with such equipment and should include a comprehensive overview of the hazards and risk from dust explosions and the control measures that are necessary to control the risk.
- Instruction to limit or control potential ignition sources.
- The importance of good housekeeping.
- The need to report any incident that may have led to a dust explosion.
- The need to report any failure of equipment, regardless of how small a failure it was.
- Awareness and training in any permit-to-work system that is installed and any safe systems of work that are used to control the risks.

LIQUEFIED PETROLEUM GAS

The term LPG refers to a group of hydrocarbon gases, two of the most common being propane and butane. Both these gases are used for many applications including the following:

- Heating
- Lighting
- Refrigeration
- Welding
- Cutting
- Cooking

Both propane and butane are odourless gases, which present their own problems as a leak cannot necessarily be detected by smell. However, commercial LPG contains a stenching agent, which is added by the manufacturers and which gives the gas a characteristic smell.

LPG has several features that are important in terms of its risk in the event of a fire. These include the following:

- LPG is heavier than air; therefore, it will sink to low levels if a leak occurs.
- LPG is highly flammable when mixed with air in concentrations between 2 and 10%.
- LPG is normally stored as a liquid in cylinders or bulk tanks.
- LPG has a large expansion ratio: it expands approximately in the ratio of 1:250 when it vaporizes from liquid to gas, therefore creating a massive gas cloud if it escapes from a cylinder.
- Large potential blast damage from an ignited gas leak, e.g. a 2.5 tonne mass of propane would produce a fireball of 40 m radius if a leak were to ignite.

Inspection of LPG Installations

LPG installations, whether in bulk tanks or in separate cylinders, need to be assessed for fire risk. A simple checklist can be used to help decide on the risk levels. The following questions should be asked:

- Siting – what are the geographic features of the site where the LPG is stored or used, e.g. slopes, where an escaped vapour may flow, or enclosed within buildings, where the vapours may get trapped.
- Location – where is the installation located, e.g. near other buildings, near boundaries of the site or near public property.
- Separation – what is the physical space between other buildings or between the periphery of the site and other properties.
- Collision protection – any device or equipment that may be installed to prevent vehicles from colliding with the installation.
- Sources of ignition – what is the potential for ignition sources being present in the area of the installation, e.g. smoking, naked flames, welding and so forth.
- Security – what arrangements are in place for securing the site or location of the installation from intruders.
- Leakage – what exists at present to prevent escapes of gas.
- Test date – all cylinders, whether they are for bulk storage or in small cylinder form, should have a test date stamped on them. This gives information of the following: date of manufacture; the relevant standard that the cylinder was made to; the test pressure used when the cylinder was tested; and the date of the last inspection.
- Procedures for dealing with emergencies, fire-fighting, dealing with leaks and spills, training of employees and other persons, and maintenance programmes for the installation.

Separation Distances

Table 9.3 gives guidance on the minimum separation distances between fixed vessels installed above ground and other buildings or risks.

Table 9.3

Minimum capacity		Minimum separation distances (m)		
(Litres)	Nominal LPG capacity (tonnes)	From buildings, property line or fixed source of ignition	With a fire wall	Between other LPG vessels
150–500	0.05–0.25	2.5	0.3	1
500–2500	0.25–1.1	3	1.5	1
2500–9000	1.1–4	7.5	4	1
9000–135 000	4–60	15	7.5	1.5
135 000–337 500	60–150	22.5	11	One quarter of the sum of the diameter of two adjacent vessels
Greater than 337 500	Greater than 150	30	15	As above

From the above table, it can be seen that separation distances can be reduced, provided a firewall has been installed. Firewalls protect vessels from thermal radiation and help to ensure adequate dispersion of leaks from vessels before they reach a boundary or a source of ignition.

Some other precautions that should be taken when LPG is stored in bulk vessels include the following:

- Prevention of collision damage by erecting crash barriers or fencing.
- Controlling sources of ignition, e.g. keeping combustible materials clear of vessels and controlling naked flames.
- Establishing a 'no smoking' policy and enforcing it strictly.
- Ensuring that the site and area around bulk vessels is secure from tampering.
- Marking with appropriate signs, e.g. 'Flammable Liquid' or 'Liquefied Petroleum Gas'.
- Ensuring that procedures are in place for filling the vessels and for maintenance work on the vessels and pipework.

LPG Cylinder Storage

LPG cylinders are used in many applications: they are used as fuels in commercial vehicles; used for heating in leisure, commercial and industrial applications and in construction sites and for lighting in various premises.

The following general precautions should be taken when storing or using LPG cylinders:

- Good general ventilation.
- Cylinders should be stored outdoors and above ground level.

- A solid base should be constructed on which to store cylinders, preferably concrete material.
- Cylinders should be stored in an upright position.
- Storage areas should not be within 3 m of openings to basements or drains.
- Cylinder storage areas should be located in secure areas and well away from site boundaries.

Table 9.4 provides general guidance for separation distances for the storage of LPG cylinders outdoors.

Storage inside a building largely depends on whether the building has been specially constructed with fire-resisting materials. Table 9.5 provides general guidance for separation within buildings.

Table 9.4

Total quantity of LPG stored (kg)	Minimum separation distance to boundary building or fixed ignition sources from the nearest cylinder (where no firewall is provided) (m)	Minimum separation distance to boundary building or fixed ignition sources from a firewall (m)
15–400	1	Nil
400–1000	3	1
1000–4000	4	1
4000–6000	5	1.5
6000–12 000	6	2
12 000–20 000	7	2.5
20 000–30 000	8	3
30 000–50 000	9	3.5
50 000–60 000	10	4
60 000–100 000	11	4.5
100 000–150 000	12	5
150 000–250 000	15	6
More than 250 000	20	7

Table 9.5

Type of building	Maximum quantity of LPG per compartment within building	Maximum number of compartments within a building	Total quantity of LPG within a building
Specially designed single-storey building.	5000 kg	5	25 000 kg
Specially designed storage space within existing buildings.	1000 kg	1	1000 kg

REFERENCES

HSE Books (1995) *Energetic and Spontaneously Combustible Substances*, HMSO.
HSE Books (1994) *Safe Handling of Combustible Dusts*, HMSO.
HSE Books (1993) *Liquefied Petroleum Gas–storage and use*, HMSO.

10 Flammable Liquids

DEFINITION OF A FLAMMABLE LIQUID

A flammable liquid is a liquid that has a flashpoint below or equal to 55 °C. The flashpoint of a substance is the 'lowest temperature at which a substance will give off sufficient flammable vapour to ignite momentarily if a flame or other ignition source is applied'. Although 55 °C is usually the cut-off point for a flammable material, it should be remembered that some substances, which do not normally fall into the category of 'flammable', should still be treated as such, especially if they are handled at temperatures above their flashpoint.

Another important category is that of substances that are 'highly flammable'. A substance categorized as highly flammable is one with a flashpoint below or equal to 32 °C. If these substances are used or stored in the workplace, a specific regulation, the Highly Flammable Liquids and Liquefied Petroleum Gases Regulations 1972, applies. These regulations require employers to put the following in place:

- Precautions to be taken during storage.
- Precautions to be taken in the event of spills and leaks.
- Controls for sources of ignition.
- Means to prevent the escape of vapours.
- Controls on smoking in the area of use or of storage.
- Dispersal of dangerous concentrations of vapours.

There are some other definitions worth mentioning regarding flammable liquids. The Chemicals (Hazard Information and Packaging for Supply) Regulations (CHIP) contain requirements for the supply of chemicals. These regulations classify flammable substances as follows:

Extremely flammable – those liquids with a flashpoint below 0 °C.
Highly flammable – those liquids with a flashpoint below 21 °C but that are not extremely flammable.
Flammable – those liquids with a flashpoint greater than or equal to 21 °C but less than or equal to 55 °C.

The above regulations require suppliers of all hazardous chemicals to:

- classify them according to their hazards;
- provide information about the hazards associated with the chemicals; and
- package the chemicals safely.

The above regulations also require suppliers to provide safety data sheets relating to the chemicals. Employers who use, store or handle extremely flammable, highly flammable or flammable substances should carry out a fire risk assessment of the activities involving those chemicals. The safety data sheets provided under the CHIP regulations provide a good source of information, which can be used to assist the task of fire risk assessment.

HAZARDS ASSOCIATED WITH FLAMMABLE LIQUIDS

A fire risk assessment carried out on the activities of using, storing or handling flammable liquids will have to identify possible hazards associated with the activities. The main hazards resulting from the use of flammable liquids are fire and explosion, which will occur if vapours are released into uncontrolled areas or ignition sources are introduced into controlled areas. Controlled areas are those areas where flammable liquids are used, stored or handled and wherein adequate precautions to prevent accidental release of vapours and introduction of ignition sources within these areas should be taken.

Common causes of accidents involving flammable liquids include the following:

- Lack of training of employees working with such substances.
- Activities involving hot work such as cutting, grinding or welding in the vicinity of flammable liquids.
- Poor design of equipment.
- Poor maintenance of equipment.
- Inadequate separation of the area where flammable liquids are used, resulting in exposure to heat from a nearby fire.
- Inadequate control of ignition sources.
- Heating substances above their self-ignition temperature.
- Inadequate arrangements and procedures during decantation from one vessel to another or from bulk tanks to smaller containers.

FIRE RISK ASSESSMENTS OF FLAMMABLE LIQUIDS

A fire risk assessment should be carried out on the activity involving flammable liquids and the likely hazards of their risk should be assessed. The following information may be required as part of a fire risk assessment:

- How are flammable liquids used or stored?
- What quantity is used or stored?
- How many people are likely to be involved in the use of flammable liquids?
- What sources of ignition are possible, e.g. naked flames, sparks, hot work, smoking materials, arson, static electricity, friction and so forth?
- Are any enclosed spaces involved?
- Where are the flammable substances used, e.g. open air or indoors?
- What type of ventilation is provided?
- What type of electrical equipment is operated nearby?
- Are vehicles operating nearby and, if so, is it possible to control them?
- What arrangements have been made for the containment of spills and leaks?
- What emergency arrangements have been made?

In addition to determining the general nature of risk, an assessment of the substance will have to be made. Information may be found in the data sheets and should include the following:

- What are the flashpoints of the substances used?
- What are the flammable limits, i.e. lower explosive limit and upper explosive limit?
- What is the viscosity of the liquids?
- What are the self-ignition temperatures of the substances?
- Are vapours heavier or lighter than air?

Most of the above definitions have been dealt with in previous chapters; however, viscosity is an important factor that has not been mentioned previously. The viscosity of a liquid determines how far any spill will spread and therefore can be used to assess any containment arrangements that are required. Solvents generally have a low viscosity and the spills will spread quickly, allowing a rapid build-up of vapours from the surface of the liquid. Paints and resins may have very high viscosities and, therefore, are easier to contain.

Another consideration is the route that escaping vapours will take. If they are heavier than air, the vapours are likely to sink to low levels, e.g. basements, drains and so forth. Vapours can travel considerable distances, depending on prevailing wind conditions. These factors are essential as far as risk assessment is concerned, as precautions may have to be taken in areas away from the immediate activity area.

Vapours that are lighter than air cause less of a problem as they generally disperse into the atmosphere. However, they are likely to get trapped in canopies or overhead structures and may present a problem as far as ignition sources are concerned.

PRECAUTIONS FOR CONTROLLING RISKS

After a fire risk assessment has been carried out, suitable and sufficient control measures should be put in place. Controls will largely depend on the nature and scale of the operation involving flammable liquids and may include some or all of the following.

Substitution

The first precaution, as always when considering the health and safety of employees, should be the elimination of the hazard. Liquids with low flashpoints should be avoided altogether and should be replaced with liquids with higher flashpoints. However, it might be that liquids that are classed as non-flammable have increased health risks, and, in this case, a balance has to be made between the elimination of one hazard and the introduction of another.

Separation

Areas where flammable liquids are used, stored or handled should normally be kept separate from other work activities. In most cases, the use of flammable liquids will involve quite small quantities, which can be safely used indoors, provided adequate precautions are taken. These will include the following:

- No smoking in the area of use.
- Keeping quantities used to a minimum.
- Ensuring that no hot work is present in the area of use.
- Adequate fire-fighting equipment nearby, e.g. foam or dry powder extinguishers.
- Adequate ventilation.

Where large quantities of flammable substances are used, it is always best to use these outside in the open air. If they are to be used indoors, special fire-resisting structures may be required to enclose the area of use, depending on the circumstances and the frequency of use.

Where flammable substances are stored in the workplace, special fire-resisting structures will be required, especially if the substances are designated as 'highly flammable'. The Highly Flammable Liquids and Liquefied Petroleum Gases Regulations include provision for the storage of highly flammable substances. However, where substances categorized as flammable, instead of highly flammable, are used, it is good practice to adhere to the requirements of the above regulations, although technically they do not apply to substances classified as flammable.

The main requirements of the regulations are as follows:

- storage should be in suitable fixed storage tanks in safe positions; or
- storage should be in suitable closed vessels, kept in a storeroom which is either in a safe place or is a fire-resisting structure; or
- storage should be in a workroom where the aggregate quantity of highly flammable liquids stored does not exceed 50 litres. Suitable closed vessels should be used to store highly flammable liquids, and these should be kept in a specially constructed fire-resisting cupboard or bin.

Cupboards and bins that meet the requirements of the regulations should have the following:

- Each side, top, floor and door of the cabinet or enclosure should be constructed of half-hour fire-resisting material.
- Each side, top, floor and door should be supported and fastened to prevent failure of the structure in a fire for at least half an hour.
- Supports and fastenings should be of non-combustible construction.
- All joints between sides, tops and floors should be bonded or fire-stopped to prevent the passage of flame and hot gases.
- The structure should be robust enough to withstand foreseeable accidental damage.
- The surface of the structure should be such that it can be easily cleaned.

Marking of Storerooms, Tanks, Cupboards and Other Storage Vessels

The regulations require that all storerooms or equipment provided for the storage of highly flammable liquids be marked with appropriate signs. The regulations require:

- clearly and boldly marked, 'HIGHLY FLAMMABLE'; or
- clearly and boldly marked, 'FLASHPOINT BELOW 32 °C'.

Where it is impracticable to mark the storage area, signs displaying the words, 'HIGHLY FLAMMABLE LIQUID' should be provided nearby, perhaps with arrows indicating the exact location of the storage area.

In addition to the above signs, 'NO SMOKING' signs should also be displayed near any storage area that is used for highly flammable liquids.

Decanting

The main hazard associated with decanting flammable liquids is that spills may occur, resulting in flammable vapours escaping. Decanting equipment should be such that the likelihood of spills occurring is minimized. Closed transfer systems are the most suitable method of transferring flammable liquids; however, depending on the results of the fire risk assessment, it may not be practical to install such an expensive option.

Specially designed small containers are available for decanting flammable liquids, which should have the following features:

- metal or heavy-duty plastic construction,
- self-closing caps,
- suitable carrying handles for larger containers,
- hoses or other devices to assist decanting into small openings,
- flame arresters on pouring or filling apertures,
- strong enough to resist dropping,
- compatible with the flammable liquid being contained,
- antistatic features to prevent the build-up of static electricity,
- ideally, meet an appropriate British or European standard.

Other precautions that may have to be adopted during decanting operations are as follows:

- Never use open topped buckets or bins for decanting into.
- Decanting should take place in well-ventilated areas, preferably outdoors.
- Decanting should never take place near areas where other flammable liquids are being used.
- Decanting should never take place near hot work. It might be necessary to stop hot work while decanting takes place.
- Drum taps should be of non-combustible material.
- Metal drums should be fitted with safety vents.

Table 10.1

Colour	Meaning or purpose	Instructions and information
Red	Prohibition sign, e.g. no smoking.	Dangerous behaviour.
	Danger alarm.	Stop, shutdown, emergency cut-out devices.
	Fire-fighting equipment.	Identification and location.
Yellow	Warning sign.	Be careful, take precautions.
Blue	Mandatory sign.	Specific behaviour or action, wear personal protective equipment.
Green	Emergency escape, first-aid sign, a place of safety.	Doors, exits, routes, emergency equipment, facilities.

- Small portable drum pumps should fit closely into the drums to minimize release of vapours.
- Electrical equipment, such as drum pumps, should be constructed to a suitable explosion-protection standard.
- All portable containers should be appropriately labelled and marked.
- Containers that have been emptied may still contain vapour and should be treated as though they were full.
- Appropriate fire-fighting equipment should be readily available and operators should be trained in their use.
- Pipes connecting tanks should ideally be situated in the open; however, where they must be situated indoors, suitable fire-resisting material should be used to insulate the pipes, especially in areas of high risk, e.g. near furnaces, hot surfaces or heating systems.
- Pipe-runs should be clearly identified and drawings provided of their locations. Special marking may be necessary where pipe-runs are located near traffic routes, vehicle access points or other areas where damage may occur.

In addition to the general precautions mentioned above, all safety signs used for marking storage areas, containers or cupboards should conform to the Health and Safety (Safety Signs and Signals) Regulations 1996. These regulations require all signs used for health and safety purposes to be designed to a specific standard.

Table 10.1 provides general guidance on safety signs:

ELECTRICAL EQUIPMENT

Electrical equipment installed in areas in which flammable liquids are used or stored should be constructed in accordance with the Electricity at Work Regulations. These regulations require employers to ensure that electrical equipment that may be exposed to any flammable substance be constructed or protected to prevent danger arising from such exposure.

Table 10.2

Zone 2	An area in which an explosive gas–air mixture is not likely to occur in normal operation, and if it occurs it will exist only for a short time. This classification can be applied if, in normal operation, there is no flammable liquid in direct contact with the surrounding atmosphere or when the area is well ventilated to disperse any flammable vapour so that any contact with electrical equipment would be only for a short time.
Zone 1	An area in which an explosive gas–air mixture is likely to occur in normal operation. This classification applies when the area contains plant and equipment that may release flammable vapours in normal operations or ventilation is such that it is not capable of dispersing vapours that may be released.
Zone 0	An area in which an explosive gas–air mixture is continuously present, or is present for long periods. This classification applies to enclosed areas that are likely to contain flammable vapour continuously or for long periods.

To comply with the above regulations, electrical equipment should be designed to prevent flammable vapours from igniting as a result of sparks or arcs from the electrical equipment. Selecting the correct equipment will depend on the circumstances in which flammable materials are used or stored. A classification scheme based on 'zones' exists to help identify the type of electrical equipment that may be used.

Hazardous areas fall into one of the three zones described in Table 10.2.

Examples of the various classifications are as follows:

Item	Extent of area	Classification
Open vessel containing flammable liquid.	Inside the vessel.	Zone 0.
Open vessel containing flammable liquid.	Vertically from ground level to 1 m above the vessel and horizontally to 2 m from the vessel.	Zone 1.
Open vessel containing flammable liquid.	Vertically to a height of 3 m and horizontally to 2 m beyond the zone 1 area.	Zone 2.

Classification of Electrical Equipment

Before electrical equipment can be used in flammable atmospheres, it should be tested and certified as suitable for use within a particular zone. There

Table 10.3

Classification of equipment	Suitable for use
Flame-proof equipment – an enclosure is used to house the electrical equipment, which, when subjected to an internal explosion, will not ignite a surrounding explosive atmosphere.	Zones 1 and 2.
Intrinsically safe equipment – a technique whereby electrical energy is limited, such that any spark or heat generated by the equipment is sufficiently low as to not ignite an explosive atmosphere.	Zone 0, 1 and 2.
Increased safety – equipment is so designed as to eliminate sparks and hot surfaces that are capable of igniting an explosive atmosphere.	Zones 1 and 2.
Non-sparking – sparking contacts are sealed against ingress of the surrounding atmosphere.	Zone 2.

are many different classifications of electrical equipment that are suitable for use in flammable atmospheres. Table 10.3 provides a summary of these classifications.

Mobile phones and radio equipment may also be capable of generating enough energy from sparks to ignite a flammable atmosphere. Arrangements should be made to eliminate the use of such equipment, especially where visitors may be present.

STATIC ELECTRICITY

The movement of liquids and gases through pipes or from decantation from one metal vessel to another can cause static electricity build-up. The most effective method of protecting against static discharge, which may result in the ignition of flammable vapours, is to ensure that all metal vessels are adequately earthed. Portable containers should also be earthed before use with bonding clips, which are used to attach containers to fixed earth points.

Depending on the findings of the fire risk assessment, it may be necessary to include the use of antistatic footwear and antistatic clothing for employees.

One important aspect of the fire risk assessment is to identify possible sources of static electricity that might not be initially obvious. For example, wire reinforcement on lagging has been the cause of several serious explosions. If the lagging is damaged and the reinforcing wire makes contact with pipes in which liquids are flowing, it is possible that electric charge may be passed through the wire and could ignite a flammable vapour if present.

It is therefore important that regular checks are carried out, not only on the earthing system but also on items of equipment or other metal components that might make contact with pipework or other apparatus.

FRICTION

Friction sparks is a common example of an ignition source that can be found in most industrial applications. Friction sparks can arise from hand tools or from operations involving rubbing or impacting. Prevention of friction includes the following controls:

- Where possible, avoid using hand tools or power tools near flammable liquids.
- Where hand tools have to be used in the vicinity of flammable liquids, spark-resisting tools should be used.
- Identify all impact-type hazards and reduce these as much as possible.
- Use permit-to-work systems, which introduce a formal control of hazards.
- Ensure that the employees working near flammable liquids are aware of the risks.

VEHICLES

Generally, special vehicle protection is not necessary where liquids with a flashpoint above 32 °C are used or stored. However, special arrangements may have to be taken if liquids with flashpoints below this are used, e.g. fitting of flame arresters.

However, even if special arrangements are not taken, drivers should be made aware of the additional risks of working in areas where flammable liquids are used or stored. In addition, if vehicles are used to transport flammable liquids, special training will have to be provided for drivers. Training should include:

- isolation of vehicles,
- first-aid fire-fighting,
- calling emergency services,
- emergency arrangements to minimize spread of flammable vapours.

HOT WORK

One of the most common causes of fires and explosions involving flammable liquids is the uncontrolled use of equipment for hot work, e.g. welding, grinding and so forth. Under no circumstances should hot work be carried out on vessels that contain or have contained flammable liquid, unless the area and the vessels have been made safe.

In order to control hot work in areas where flammable liquids or vapours exist, and reduce human failure, which can have catastrophic effects, the use of a permit-to-work system would be mandatory. Permits are systems that ensure that formal action is taken to control risks by providing written and signed statements confirming that action has been taken.

Permits should be the overall responsibility of one person. This person should be competent to recognize fire hazards associated with flammable liquids and should have the authority to be able to put in place an action that is considered necessary.

Design of a Permit-to-Work Form

A typical permit-to-work form should include the following information:

- The signature of the manager, releasing the plant or equipment for the task.
- Details of any isolation necessary for the task to be completed safely. Isolations should have a signature confirming that such isolation has taken place.
- Signature of the person carrying out the task to confirm that any other precautions that are necessary have been taken, e.g. fire extinguishers standing by, area dampened down before cutting work commences, purging of vessels or containers before work commences, elimination of electrical equipment.
- Signature of a supervisor or team leader to confirm that all necessary precautions have been taken to ensure that the task can be carried out safely.
- Signature of the above person to confirm that the task has been completed successfully or that the task has been abandoned, in which case the permit should be cancelled and another one started when the task recommences.

There are four main principles of operation of permit-to-work systems. These are discussed in the following sections.

Hazard Evaluation

This stage involves a detailed evaluation of all hazards that are likely to be present at the start and during the task. This stage should normally be carried out by a number of persons, e.g. supervisors or managers, the person carrying out the task and any specialist, e.g. engineers.

Planning of Precautions

A competent person should carry out planning before a task is started and before the permit is issued. Planning is a method of ensuring that the hazards identified in the first stage can be controlled for their risk. If some of the hazards cannot be eliminated or controlled, suitable alternative arrangements should be made.

Instructing the Supervisors

Any person who is responsible for the work, e.g. a supervisor or manager, should be made aware of the planning that has gone into the precautions taken, especially if this person has not been part of the planning stage. The person responsible for issuing and checking that the permit remains in force should be of sufficient status to be able to stop work at any time should he or she deem the situation to be hazardous.

Issuing the Permit

The permit, which should be signed by the issuer, should be given to the person carrying out the task, and sufficient copies should be given to other key personnel, e.g. engineers, managers and so forth. A copy of the permit, an example of which is shown below, should be displayed near the task being carried out so that anyone entering that area can clearly see it.

Example of a Permit-to-Work Form

Plant details	Description of the equipment to be worked on and the quantity and description of any flammable liquids nearby.
Details of the work to be carried out	Description of the task and the type of operation involved, e.g. welding, cutting, grinding. This should include the number of people likely to be carrying out the task and the anticipated duration of the task.
Withdrawal from service	Details of any equipment that has to be withdrawn from service during the work or of any restrictions made to the area, e.g. vehicles or pedestrians.
Isolations	Details of any isolations that are necessary during the task, e.g. flow pipes entering the flammable liquid vessel or electrical equipment that has to be turned off. Isolations should be signed for.
Cleaning and purging	Any purging or cleaning operation that is necessary before the work commences should be detailed and signed for.
Testing	Details of any gas testing should be recorded. This should include the type of equipment used to carry out the test and the person who carried out the test.
Special arrangements	Details of any additional arrangements, e.g. fire extinguishers to be positioned nearby, fire-watchers to be stationed in the area, fire alarms to be isolated to prevent accidental actuation.
Acceptance	This section should include a signature of the person accepting responsibility for the task and for the person carrying out the task.
Completion of work	This section should include a signature to 'sign-off' the permit after the work has been completed.
Return to service	Finally, any isolations or withdrawn equipment should be reinstated and signed for.

In addition to the permit-to-work system, the following precautions should be taken:

- Flammable liquid should be drained from vessels or containers before work commences.
- The vessel should be inspected before work commences to ensure that all residues have been removed.
- All supplies to the vessel or container should be isolated to ensure that flammable liquids cannot inadvertently enter the item while work is in progress.
- The vessel should be inerted with, for example, water, nitrogen or foam. This will ensure that any unseen vapours are removed before work commences.
- Gas tests should be carried out to monitor the contents of the container and the surrounding area. Gas tests may have to be continued throughout the work operation.

SMOKING

Under no circumstances should employees or others be allowed to smoke while in the area of flammable liquids. 'No smoking' signs should be displayed around any area that is used for storing or using flammable liquids, and all employees should be instructed to extinguish cigarettes before entering these areas.

Special care should be taken while visitors or contractors are present in these areas. A nominated person from the company should be responsible for visitors and contractors and should notify them of any area designated as a 'no smoking' zone.

VENTILATION

Good ventilation is essential in areas where flammable liquids are used or stored. As far as possible, flammable materials should be used in the open. However, where this is not practical, special booths may be required for work involving flammable liquids.

If such areas are used, adequate ventilation should be installed and should aim to reduce any flammable vapour below 25% of its lower explosive limit. Ventilation systems should exhaust to open air, well away from vehicles or other operations involving sparks or naked flames. As a general rule, openings for venting flammable vapours should be at a height of at least 3 m above ground level and at least 3 m from building and basement openings.

Ventilation systems used for extracting flammable vapours should be tested by a competent engineer, at least annually. Ducting should be checked for its integrity, and flow rates should be taken to ensure that suction is adequate.

Depending on the type of flammable vapour, it may be necessary to install an airflow detector and alarm system. This would alert employees that the suction or airflow had failed, enabling the system to be shut off. Where the fire risk assessment suggests that a high fire risk is present, it might be necessary to install automatic shut-off systems.

HOUSEKEEPING

As always, housekeeping is extremely important to prevent fires. As far as flammable liquids are concerned, the following precautions should be taken:

- Leaks and spills of flammable liquids should be dealt with promptly.
- A supply of absorbent material should be made available in areas where flammable liquids are used or stored.
- Suitable receptacles should also be provided for disposing of absorbent material that has become contaminated with liquids.
- Drains around areas where flammable liquids are used should be fitted with interceptors to prevent spilled liquids from entering the main drainage system.
- Suitable receptacles should be provided for rags and cloths to be disposed in. These should be fitted with self-closing lids and should not be used for general waste material.
- Bins should be emptied on a daily basis.
- Bins used for storing rags and cloths should not be positioned in areas of direct sunlight.
- Bins used for storing rags or for the disposal of absorbent material should be suitably marked e.g. solvent waste.

TRAINING

All employees who work with flammable liquids and who are responsible for storing these materials should be provided with adequate information and training on the risks of fire and explosion. Information and training should include the following:

- Types of flammable materials used.
- Where to obtain information about the materials used, e.g. COSHH Data Sheets, which should be located near the area where the flammable materials are used or stored.
- General procedures for safe handling of flammable liquids.
- The correct use of personal protective clothing.
- Dealing with emergencies, e.g. fires, spills.
- First-aid fire-fighting and the use of fire-fighting equipment.
- Shutting down procedures for plant or equipment.
- Evacuating the area.

Table 10.4

Tank capacity (m^3)	Separation distance (m)
Less than 1	2
Between 1 and 5	4
Between 5 and 33	6
Between 33 and 100	8
Between 100 and 250	10
Greater than 250	15

STORAGE OF FLAMMABLE LIQUIDS IN FIXED TANKS

Although the precautions discussed so far apply to both portable containers and small tanks, they apply equally to larger, fixed tanks. However, fixed tanks require some further considerations.

Separation Distances of Small Fixed Tanks

A small tank is generally taken to be one with a diameter of less than 10 m. The minimum separation distances given in Table 10.4 relate to distances between the fixed tank and any of the following:

- Building
- Boundary of premises
- Process unit
- Fixed ignition source
- Any door, internal or external

It is acceptable to group small tanks together in one area. A tank is considered as part of a group if adjacent tanks are within the separating distances given in Table 10.4; however, the aggregate capacity of the group of tanks should be no more than 8000 m^3.

If tanks are grouped together, consideration should be given to the following:

- Adequate space between the tanks to allow safe access and egress.
- Safe distances to allow movement of people in emergency situations.
- Adequate means of containing spills or leaks.
- Adequate means of fire-fighting, e.g. enough fire extinguishers.

Separation Distances for Large Tanks

For larger tanks, the separation distance depends on the type of tank used to store the flammable liquid. Tanks can either be of fixed roof construction or floating roof construction. See Table 10.5.

Table 10.5

Factor	Minimum separation from any part of the tank
Between adjacent fixed-roof tanks	Equal to the smaller of the following: • the diameter of the smaller tank; • half the diameter of the larger tank; • 15 m.
Between adjacent floating-roof tanks	10 m for tanks up to and including 45 m diameter. 15 m for tanks over 45 m diameter.
Between a floating-roof and a fixed-roof tank	Equal to the smaller of the following: • the diameter of the smaller tank; • half the diameter of the larger tank; • 15 m; but not less than 10 m.
Between a group of small tanks and any tank outside that group.	15 m.
Between a tank and the site boundary, any designated non-hazardous area, process area or any fixed source of ignition.	15 m.

Storage of Flammable Liquids Inside Buildings

Flammable liquids should not normally be stored inside buildings in fixed tanks. If storage is necessary, small quantities only should be stored and even these should be stored for minimum periods, e.g. the duration of a working day or a shift.

Where it is essential that flammable liquids are stored inside buildings, suitable fire risk assessments should be carried out and suitable control measures should be put in place. Suitable control measures may include the following:

- Storage in a single-storey building, which is of non-combustible construction.
- A suitable means of explosion relief installed within the roof of the building.
- Mechanical ventilation to ensure that explosive atmospheres are not generated should a vapour release occur.
- A very high standard of general ventilation.
- Fire separation between the part of the building housing the tank and any other part of the building.
- Fire separation of 4 m between the building used to store the tank and any other building.
- Adequate means of raising the alarm in the event of a fire.

- Adequate means of fire-fighting, e.g. large quantities of foam extinguishers.
- Adequate means of escape.

LOADING AND UNLOADING OF TANKERS

Loading and unloading of fixed tanks by road tankers has resulted in many fires and explosions. Usually, these have occurred because of poor general precautions not being adhered to. During such operations, the following general precautions should be adopted:

- Loading and unloading bays should be located in safe, well-ventilated areas, which are well away from other buildings or hazards, e.g. a minimum of 10 m from buildings and boundaries.
- Easy access should be allowed for tankers and other vehicles.
- Adequate lighting to the appropriate standard of electrical protection should be installed.
- The use of flow meters with trip cut-out devices to prevent overfilling.
- The use of breakaway couplings to prevent tankers being driven away while still connected to the supply.
- Suitable earthing to prevent static electricity.
- Adequate fire-fighting arrangements to contain any free-flowing liquid fire, i.e. foam equipment.

REFERENCES

HSE Books (1996) *The Safe Use and Handling of Flammable Liquids*, HMSO.
HSE Books (1998) *The Storage of Flammable Liquids in Tanks*, HMSO.

Index